POPULAR LECTURES IN MATHEMATICS

Editors: I. N. SNEDDON and M. STARK

VOLUME 10

Mathematical Games
and·Pastimes

Mathematical Games and Pastimes

A. P. DOMORYAD

TRANSLATED BY

HALINA MOSS

A Pergamon Press Book

THE MACMILLAN COMPANY

NEW YORK

1964

THE MACMILLAN COMPANY
60 Fifth Avenue
New York 11, N.Y.

This book is distributed by
THE MACMILLAN COMPANY
pursuant to a special arrangement with
PERGAMON PRESS LIMITED
Oxford, England

This is a translation from the original Russian
Matematicheskiye igry i razvlecheniya published
in 1961 by Fizmatgiz, Moscow

Library of Congress Catalog Card Number 63–16860

MADE IN GREAT BRITAIN

Contents

Contents

Foreword

FROM the wide variety of material collected by various authors under the name of mathematical games and pastimes, there can be extracted several groups of "classical pastimes", which drew the attention of mathematicians for a long time;

(1) Pastimes, connected with the search for original solutions of problems, which permit a practically unlimited number of solutions (see e. g. "Magic squares" — ch. 16, "The problem of the chess knight" — ch. 19, etc.). Here, the interest is usually centred on establishing how many solutions there are, working out methods leading to large groups of solutions, or on solutions satisfying some special requirements.

(2) Mathematical games, i. e. games in which two players aim at a definite goal, through a number of "moves" made one after the other in accordance with agreed rules: here it turns out to be possible to predetermine the victor for any initial situation, and to indicate how he is to win, no matter what the opponent's moves are (see e. g. ch. 10).

(3) "Games for one person", i. e. pastimes in which it is necessary to reach a definite, predetermined goal by means of a number of operations, carried out by the player himself, in accordance with given rules (see e.g. chs. 11–14): here the interest is centred on the conditions under which the goal is reached, and it is required to find the least number of moves necessary to reach that goal.

The greater part of this book is devoted to classical games.

The first few chapters deal with various systems of notation and with certain topics in the theory of numbers, the knowledge of which is necessary for the understanding of the theory of various mathematical games. But for some readers these chapters might be interesting in themselves.

The theory of some isolated games is presented fairly fully here; in other cases only results are given; and reference is made to sources, where proof of these results can be found.

Side by side with classical pastimes, the book devotes much space also to "contemporary" pastimes quick reckoning, re-cutting of figures, construction of curves, and models of polyhedra.

Deserving particular attention are the problems which admit a practically inexhaustible or even infinite number of solutions ("Construction of parquets", "Construction of pleasing patterns", etc.).

Here, everybody, by applying persistence and inventiveness, can attempt to obtain interesting results.

Whereas such classical pastimes as, for example, constructing "magic squares" may be enjoyed by a comparatively narrow section of people, the cutting out of, say, symmetrical figures in paper, the construction of pleasing patterns, searching for numerical curiosities, by not requiring any mathematical preparation, might give pleasure to both amateur and professional mathematicians. The same can be said about pastimes requiring knowledge confined to that obtained in the 8th to 10th classes of the secondary school (construction of parquets, of interesting curves and borders, etc.).

In group activities it is possible to arrange competitions in making up original parquets, in the construction of curves and borders, in obtaining attractive symmetrical figures cut out of paper, and so on. Each participant in such competitions can dazzle with his inventiveness, accuracy of execution, or artistry of colouring the figures obtained.

Such collective activity can be rounded off by com-

piling an album or by organizing an exhibition of the
best items.

Many pastimes and even single problems may suggest
to the amateur mathematician themes for independent
investigations (the use of knight's moves instead of the
"short" moves of the fook in the "game of 15", the
search for interesting identities — see § 37 —, the generalization of the problem about tourists — problem
No. 13 in § 37 — and so on).

On the whole, this book caters for readers with
mathematical knowledge within the limits of the 9th
and 10th classes of the secondary school, even though
the greatest part of the material is accessible to pupils
of the 8th class, and some topics — even to schoolchildren of the 5th and 6th classes.

Many chapters can be used by teachers of mathematics for extracurricular activities.

Various categories of readers can use the book in
various ways: persons not particularly fond of mathematics can become acquainted with curious properties
of numbers or figures, without going into the funda-
mentals of the games and pastimes, and taking for
granted single propositions; amateur mathematicians
are advised to study certain parts of the book with
pencil and paper, solving the problems given and
answering the questions posed.

§ 38 gives answers to the problems to be found in
the text, questions and hints towards their solution
and also proofs of certain of the theorems mentioned in
the text. References to the appropriate section of § 38
are given in small figures between ordinary brackets.

References to books in which the reader may find a
more detailed discussion of the topics touched upon are
given by a number enclosed in square brackets. This
number refers to the corresponding entry in the biblio-
graphy at the end of the book.

§ 1. Various systems of notation

A certain amateur mathematician had the following notes in his jotter:

$$\begin{array}{r} 3205 \\ +\,4775 \\ \hline 10202 \end{array}$$ (five and five is ten: we write 2 and carry 1, and so on . . .)

$$\begin{array}{r} 3217 \\ -\,1452 \\ \hline 1545 \end{array}$$

$$\begin{array}{r} 435 \\ \times\;47 \\ \hline 3713 \\ 2164 \\ \hline 25553 \end{array}$$ (five sevens are thirty-five: we write 3 and carry 4 and so on . . .)

$$\sqrt{104231} = 273$$

$$\begin{array}{r} 4 \\ \hline 47\ |442 \\ \times\ 7\ |421 \\ \hline 563\ |2131 \\ \times\ 3\ |2131 \\ \hline 0 \end{array}$$

$$\frac{17}{43} = \frac{3}{7} \quad \text{(cancelling by five)}$$

$$\begin{array}{r} 361 \\ 30 \\ \hline 61 \\ 60 \\ \hline 100 \\ 74 \\ \hline 40 \\ 30 \\ \hline 100 \end{array}$$ and so on

$$\begin{array}{l} 14 \\ \hline 24\cdot 0525 \end{array}$$

Therefore $\dfrac{361}{14} = 24\cdot 052$

Verification: $0\cdot 052 = \dfrac{52}{770} = \dfrac{1}{14}$

(after cancelling by 52);

$$24 + \frac{1}{14} = \frac{24 \times 14 + 1}{14} = \frac{361}{14}$$

At first glance all these operations make a very queer impression; however, everything becomes clear when it is taken into account that all operations were carried out in the system of notation with the base 8.

The crux of the matter is that in our usual system of notation, the separate digits of the number N, depending on their place, indicate the number of units, tens, hundreds, etc., or the number of tenth, hundredth, etc., parts forming the number N.

By selecting, as the base of a system of notation, any number k, i. e., by regarding k units of any order (and not ten units, as is done in the decimal system) as forming one unit of the next largest order, we arrive at the so-called system of counting to the base k.

If $k<10$, the digits from k to nine become superfluous (it is no accident that in all examples cited above the digits 8 and 9 are absent!). If $k>10$ then it is necessary to invent symbols for numbers from 10 to $k-1$ inclusive; for example, in the duodecimal system, the numbers 10 and 11 may be denoted by α and β respectively.

When a number is written down in the base k system, it is convenient to indicate in brackets (on the right, below) the base of the system as written down in the customary decimal system, for example

$$
\left.
\begin{aligned}
1\ 101_{(2)} &= 1\times2^3+1\times2^2+0\times2+1 = 13, \\
20\ 120_{(3)} &= 2\times3^4+1\times3^2+2\times3 = 177, \\
\alpha13\beta_{(12)} &= 10\times12^3+1\times12^2+3\times12+11 = 17471, \\
1\cdot672_{(8)} &= 1+\frac{6}{8}+\frac{7}{8^2}+\frac{2}{8^3}=\frac{477}{256}
\end{aligned}
\right\} \quad (1)
$$

In the last example on the left, there is a so-called *r a d i x   f r a c t i o n*, which is analogous to a decimal fraction.

When k is large, the numbers 10 to $k-1$ can be written down by making use of the decimal system, joining these number-symbols by a short stroke on top; for example:

$$
\left.
\begin{aligned}
\overline{10}\ 0\ 6\ \overline{11}_{(16)} &= 10\times16^3+6\times16+11 = 4203, \\
3\ \overline{13}\ \overline{12}\ \overline{41}_{(60)} &= 3\times60^3+13\times60^2+12\times60+ \\
&\quad +41 = 695561, \\
0\cdot\overline{30}\ \overline{10}_{(60)} &= \frac{30}{60}+\frac{10}{60^2}=\frac{181}{360}
\end{aligned}
\right\} \quad (2)
$$

It can be seen from equalities in (1) and (2) that it is quite simple to change over from writing a number down in the base-k system to writing it down in the convenient decimal system.

It is also quite easy to solve a converse problem: to write down a natural number N given in the decimal system, in the system with base k.

Let
$$
\begin{aligned}
N &= kq_1 + c_0, \\
q_1 &= kq_2 + c_1, \\
q_2 &= kq_3 + c_2, \\
&\cdots\cdots\cdots \\
q_{n-2} &= kq_{n-1} + c_{n-2}, \\
q_{n-1} &= kq_n + c_{n-1}.
\end{aligned}
$$

Here q_1 and c_0 are the quotient and the remainder obtained, when N is divided by k; in general q_{s+1} and c_s are the quotient and remainder obtained in the division of q_s by k. Each of the remainders $c_0, c_1, c_2, \ldots,$ c_{n-1} is less than k, but greater than or equal to zero; $0 < q_n < k$. Hence

$$
\begin{aligned}
N &= (kq_2 + c_1)\,k + c_0 = q_2k^2 + c_1k + c_0 = \\
&= (q_3k + c_2)\,k^2 + c_1k + c_0 = \ldots = q_nk^n + c_{n-1}k^{n-1} + \\
&+ c_{n-2}k^{n-2} + \ldots + c_2k^2 + c_1k + c_0 = \\
&= q_nc_{n-1}c_{n-2}\ldots c_2c_1c_{0(k)}.
\end{aligned}
$$

It is convenient to carry out calculations according to a scheme, which becomes clear from the following example; *express* 695561 *in the system of notation with the base sixty.*

$$
\begin{array}{r|r|r|r|r}
695561 & 60 & & & \\
\underline{60} & \underline{11592} & 60 & & \\
95 & \underline{60} & \underline{193} & 60 & \\
\underline{60} & 559 & \underline{180} & \underline{3} & \\
355 & \underline{540} & 13 & & \\
\underline{300} & 192 & & & \\
556 & \underline{180} & & & \\
\underline{540} & 12 & & & \\
161 & & & & \\
\underline{120} & & & & \\
41 & & & &
\end{array}
$$

Therefore

$$695561 = 3 \ \overline{13} \ \overline{12} \ \overline{41}_{(60)}.$$

If $N = \frac{a}{b}$, then in order to write N down in the system of notation with base k, it is sufficient to write down a and b in this system. The fraction obtained can be represented in the form of a base-k fraction, by dividing a by b in the base-k system. Here the number a (we suppose $a < b$) and the remainders obtained in the process of division, must be split up into units of lower orders $\left(\text{equal } \frac{1}{k}, \frac{1}{k^2}, \frac{1}{k^3}, \ldots\right)$ adding on a zero on the right-hand side of each of them. Let us represent, for instance, $\frac{17}{18}$ in the duodecimal system, and $\frac{4}{7}$ in the system with base three:

$$\frac{17}{18} = \frac{15_{(12)}}{16_{(12)}} = 0 \cdot \overline{114}_{(12)}, \text{ since when } k = 12$$

$$
\begin{array}{r|l}
15 \cdot 0 & 16 \\
\underline{14\ 6} & 0 \cdot \overline{11}\ 4 \\
60 & \\
\underline{60} & \\
0 &
\end{array}
$$

$$\frac{4}{7} = \frac{11_{(3)}}{21_{(3)}} = 0 \cdot (120102)_{(3)}, \text{ since}$$

$$
\begin{array}{r|l}
11 \cdot 0 & 21 \\
\underline{21} & 0 \cdot 120102\ldots \\
120 & \\
\underline{112} & \\
100 & \\
\underline{21} & \\
200 & \\
\underline{112} & \\
11 \text{ and so on} &
\end{array}
$$

(the last remainder $11_{(3)}$ coincides with the initial number $a = 11_{(3)}$, therefore, the required fraction is *recurring*.

4

It is easy to verify the truth of the result obtained by applying the following rule: in order to convert a recurring base-k fraction, into a vulgar fraction, the whole period of the fraction should be divided by a number, which is written down by means of as many "$k-1$" as there are digits in the period (prove ([1]) this rule, making use of the formula for the sum of terms of an infinitely decreasing geometrical progression).

In this case $k = 3$, therefore

$$0 \cdot (120102)_{(3)} = \frac{120102_{(3)}}{222222_{(3)}} - \frac{11_{(3)}}{21_{(3)}}$$

(after cancelling by $10212_{(3)}$ — check!).

In order to make it easy to multiply and divide numbers in the system with base-k, it is useful to have multiplication tables, which give products of pairs of numbers not exceeding $k-1$.

For example, for $k = 8$ and for $k = 12$ we have respectively:

	2	3	4	5	6	7		2	3	4	5	6	7	8	9	α	β
2	4	6	10	12	14	16	2	4	6	8	α	10	12	14	16	18	1α
3	6	11	14	17	22	25	3	6	9	10	13	16	19	20	23	26	29
4	10	14	20	24	30	34	4	8	10	14	18	20	24	28	30	34	38
5	12	17	24	31	36	43	5	α	13	18	21	26	2β	34	39	42	47
6	14	22	30	36	44	52	6	10	16	20	26	30	36	40	46	50	56
7	16	25	34	43	52	61	7	12	19	24	2β	36	41	48	53	5α	65
							8	14	20	28	34	40	48	54	60	68	74
							9	16	23	30	39	46	53	60	69	76	83
							α	18	26	34	42	50	5α	68	76	84	92
							β	1α	29	38	47	56	65	74	83	92	$\alpha1$

For example, $5 \times 6 = 36_{(8)}$; $5 \times 7 - 2\beta_{(12)}$.

Prove ([2]), that the extraction of square roots in any system of notation is carried out exactly in the same way as in the decimal system (see example at the beginning of this chapter).

The Binary System of Notation

In the binary system of notation any whole number can be written down by means of the digits 1 and 0; this means, that any natural number is a sum of various powers of 2:

$$N = 2^{\alpha_1}+2^{\alpha_2}+ \ldots +2^{\alpha_s} \ (\alpha_1 > \alpha_2 > \ldots > \alpha_s \geqslant 0).$$

The trick involving guessing a number is based on this property of integers: on cards with "headings" 1, 2, 4, 8, 16 (Fig. 1), integers are written down in such a way that any given number N occurs only on those cards the sum of whose headings equals N. For example, 27 $(1+2+8+16)$ should be absent only on the card with the heading 4 and so on.

1	17		2	18		4	20		8	24		16	24
3	19		3	19		5	21		9	25		17	25
5	21		6	22		6	22		10	26		18	26
7	23		7	23		7	23		11	27		19	27
9	25		10	26		12	28		12	28		20	28
11	27		11	27		13	29		13	29		21	29
13	29		14	30		14	30		14	30		22	30
15	31		15	31		15	31		15	31		23	31

Fig. 1.

Having got someone to think of a number not exceeding 31, and to point out exactly in which cards the number is to be found, it is possible to guess immediately what that number was, by adding up the numbers in the headings of the cards pointed out.

This trick can be mechanized by writing out the tables indicated on the laminae which weigh 1, 2, 4, 8, 16 g respectively. If the laminae containing the number thought of are placed on a sufficiently sensitive spring balance, the pointer will come to rest at that number. Another possibility of mechanizing the trick is referred to in [25], p. 71.

The binary system of notation is used frequently in contemporary electronic computers. The situation is that elements, which are used to represent numbers in these machines, can exist in two easily distinguishable states (for example, positive and negative charge of a portion of the dielectric, opposite magnetization of a portion of magnetic tape, etc.). Thus, each such element can be used to represent one order of the number expressed in the binary system (one state represents zero in a given order, the other state represents 1). It is also of some significance, that it is much simpler to operate on two kinds of digits (0 and 1) only. The binary system finds some application also in the theory of games with three piles of objects (see § 10).

The System of Notation with the Base Three

In the system of notation with the base 3, any whole number can be represented by means of the digits 0, 1 and 2. However, if "negative digits" are introduced, as is done, for example, in representing logarithms with a negative characteristic, then it follows from the equation

$$2 \times 3^m = 3^{m+1} - 3^m = 1 \times 3^{m+1} + \bar{1} \times 3^m$$

that any number can be represented in the system with base 3 by means of the digits 0, 1, $\bar{1}$ and therefore, we have the following.

THEOREM. *Any integer is an algebraic sum of various powers of three, i. e.*

$$N = 3^{\alpha_1} + 3^{\alpha_2} + \ldots - 3^{\beta_1} - 3^{\beta_2} - \ldots, \qquad (3)$$

where α_1, α_2, $\ldots$, β_1, β_2, $\ldots$ are various non-negative integers; there might not necessarily be any negative terms in the eqn (3).

For instance, for the number 1910 we have:

$N = 1910 = 2121202_{(3)} = 21212 1\bar{1}_{(3)} = 2122\bar{1}\bar{1}\bar{1}_{(3)} =$
$= 213\bar{1}11\bar{1}_{(3)} = 220\bar{1}111_{(3)} = \bar{3}10\bar{1}111_{(3)} = 10\bar{1}0\bar{1}111_{(3)}$
$= 3^7 - 3^5 - 3^3 - 3^2 + 3 - 3^0$

(in the process of transformation, the 3 acts temporarily as a digit).

Hence there follows naturally the solution of the ancient problem about four weights, by means of which any load from 1 to 40 kg can be weighed on a beam balance†).

For a detailed analysis of the problem and its generalization, see [30], pp. 176–178 (supplemented by S. O. Shatunovsky) and in [25], pp. 76–90.

Indeed, by placing on one pan of the balance weights of 3^{α_1} kg, 3^{α_2} kg, etc., and on the other pan — 3^{β_1} kg, 3^{β_2} kg, etc. [see (3)], we can weigh a load of N kg. Therefore a set of weights of 1 kg, 3 kg, 9 kg, ..., 3^n kg permits the weighing of any whole-number load of N kg, where

$$N \leqslant 1 + 3 + 9 + \ldots + 3^n - \frac{3^{n+1} - 1}{2}$$

when $n = 3$

$$\frac{3^{n+1} - 1}{2} = 40$$

PROBLEMS. 1. Represent ([3a]) the numbers 2713 and 409 in the system of notation with the base 5, by means of the digits 0, 1, 2 (you are allowed to use "negative" digits).

2. Verify the correctness of the operations carried out at the beginning of this chapter in two ways:

(a) by carrying out all operations directly in the scale to base 8,

(b) by representing the numbers with which the operations were carried out, and the resulting numbers, in the decimal system.

3. Represent $\frac{1}{7}$ and $\frac{1}{16}$ in the form of fractions in the binary, base-3, duodecimal, and base-60 systems. For verification, transform the fractions obtained into ordinary ones.

4. Having represented 676 in the binary, base-3 and base-5 systems, extract the square root of the numbers obtained.

5. Show ([3b]) that in order to change over from the base-8 system to the binary one it is sufficient to express each figure as a three-digit number in the binary system; for example:

$$7315_{(8)} = \underbrace{111}\,\underbrace{011}\,\underbrace{001}\,\underbrace{101}_{(2)}$$

and conversely

$$\underbrace{10}\,\underbrace{000}\,\underbrace{101}\,\underbrace{110}_{(2)} = 2056_{(8)}.$$

†) Footnote: the problem of the weights was investigated as far back as A. D. 1202 by Leonardo of Pisa (Fibonacci).

Various Systems of Notation

Taking into account this rule and analogous rules for the transition from the system to the base 4 and the system to the base 16 to the binary one, etc., show that

$$\overline{11}\ 4\ \overline{13}_{(16)} = 5515_{(8)} \text{ and } 773_{(8)} = 13323_{(4)}.$$

6. Prove ([4]) that when $k > 5$ the number $123454321_{(k)}$ is a perfect square.

7. How is it possible ([5]) to determine the natural number N (up to 1000) thought of by someone, by asking 10 questions the answer to each of which is "yes" or "no"?

8. It is no accident, that each of the cards in Fig. 1. contains 16 numbers.

In general, if s cards with headings $1, 2, 4, 8, \ldots, 2^{s-2}, 2^{s-1}$ are taken, and any number m (from 1 to 2^{s-1}) is written in all those cards, whose headings add up to m, then each card will contain 2^{s-1} numbers. Attempt to prove this theorem([6]).

§ 2. Some Facts from the Theory of Numbers

If, given a and b, it is possible to pick such a number c, that $a = bc$, it is said that a is divisible by b, and b is called the *divisor* of the number a (unless otherwise stated, only natural numbers are dealt with in this chapter).

The number p is called *prime* if it has two positive divisors only; 1 and p.

Any composite (i. e. not prime) number n can be represented in the form:

$$n = p_1^\alpha p_2^\beta \ldots p_k^\sigma, \tag{1}$$

where $p_1, p_2, \ldots, p_k$ are prime numbers and $\alpha, \beta, \ldots, \sigma$ are natural numbers: if there are no identical numbers among $p_1, p_2, \ldots, p_k$, then (1) is called the *canonical* factorization of n.

In any course on the subject of the theory of numbers there is to be found the proof of the following

THEOREM. *For any number n there exists only one canonical factorization* (provided factorisations differing only in the order of factors are not regarded as different).

Some readers might regard it as strange that mathematicians waste efforts on proving such self-evident theorems. However, the following example, taken from another numerical region (not from the realm of natural numbers) shows that a perfectly analogous and seemingly self-evident theorem can be wrong. We shall call "composite" the complex number $a + b\sqrt{(-6)}$, where a and b are any whole numbers, if it can be represented in the form:

$$a + b\sqrt{-6} = (c + d\sqrt{-6})(e + f\sqrt{-6}),$$

where c, d, e and f are whole numbers (which can, in particular, be zeros) and each factor is neither 1 nor —1. Otherwise we call it prime.

According to this definition, the numbers $20 - \sqrt{(-6)}$, 7 (here $a = 7$, $b = 0$), and 6 ($a = 6$, $b = 0$) are composite numbers:

$$20 - \sqrt{-6} = (2+3\sqrt{-6})(1-2\sqrt{-6}),$$
$$7 = (1+\sqrt{-6})(1-\sqrt{-6}),$$
$$6 = 2 \times 3 - (\sqrt{-6}) \times (-\sqrt{-6}).$$

But it is possible to prove, that numbers $\sqrt{(-6)}$, $-\sqrt{(-6)}$, 2, 3 are "prime" numbers (see e. g. [24], pp. 84—85). It follows that the composite number 6 can be factorized into primes in two different ways!

Functions $\tau(n)$ and $S(n)$

Let us denote the number of positive divisors of n by $\tau(n)$, and their sum by $S(n)$.

For example, $\tau(10) = 4$ and $S(10) = 18$, since 10 has only four positive divisors; 1, 2, 5, 10.

If $n - p_1^a p_2^\beta \ldots p_k$ is the canonical factorization of n, then

$$\tau(n) = (\alpha+1)(\beta+1)(\gamma+1)\ldots(\sigma+1), \qquad (2)$$
$$S(n) = (1+p_1+p_1^2+\ldots+p_1^a)(1+p_2+\ldots+p_2^\beta) \times$$
$$\times (1+p_3+\ldots+p_3^\gamma)\ldots(1+p_k+\ldots+p_k^g). \quad (3)$$

Indeed, any number of the form

$$p_1^{a'} p_2^{\beta'} p_3^{\gamma'} \ldots p^{\sigma'}, \qquad (4)$$

in which

$$0 \leqslant \alpha' \leqslant \alpha; \ 0 \leqslant \beta' \leqslant \beta; \ 0 \leqslant \gamma \leqslant \gamma; \ \ldots; \ 0 \leqslant \sigma' \leqslant \sigma. \qquad (5)$$

is a divisor of n.

Since α' can be selected in $(\alpha + 1)$ ways, β' can be selected in $(\beta + 1)$ ways, etc., there exist $(\alpha + 1)$ $(\beta + 1)$ ways of selecting a pair of numbers α', β' (each specific value α' can be combined with any of the $\beta + 1$ values of β'), and there are $(\alpha + 1)(\beta + 1)$ $(\gamma + 1)$ ways of selecting a triplet of numbers α', β', γ' (any specific pair α', β' can be combined with any of the $\gamma + 1$ values of γ').

Continuing this reasoning, we conclude that the group of numbers α', β', γ', ..., σ', satisfying condition (5) can be selected in $(\alpha + 1)\,(\beta + 1)\,(\gamma + 1)\ldots(\sigma + 1)$ ways. Therefore formula (2) is correct. The truth of formula (3) is deduced from the fact that on multiplying out the polynomials in the right-hand-side of (3) we obtain the sum of all possible terms of form (4).

If $S(n)$ - $2n$, the number n is called *perfect*. For example, 6 and 28 are perfect numbers, since $S\,(6) = 12$ and $S(28) = 56$.

Euclid established that even numbers of the form

$$N = 2^{\alpha}\,(2^{\alpha+1} - 1), \tag{6}$$

where α is a natural number and $2^{\alpha+1} - 1$ is a prime number are perfect (see [2], pp. 72–74). On the other hand, no perfect even numbers exist with any other canonical factorization.

So far, it is established that numbers of the form $2^{\alpha+1} - 1$ (called Mersenne numbers, after a French scholar of the XVII century) are prime for $\alpha = 1, 2, 4, 6, 12, 16, 30, 60, 88, 106, 126, 520, 606, 1278, 2202, 2280$; that is, so far only seventeen perfect numbers are known.

The first seven perfect numbers are as follows

6, 28, 496, 8128, 33 550 336, 8 589 869 056, 137 438 691 328.

To this day it is not known whether there exist odd perfect numbers.

In mediaeval times, mystics among mathematicians paid great attention to so-called amicable numbers, that is numbers a and b, for which $S(a) = S(b) = a + b$. Using formula (3), prove that 220 and 284 are amicable numbers. Numbers 18416 and 17246 are also amicable.

Function x (the integral part of x)

The function $[x]$ equals the greatest integer not exceeding x (x is any real number).

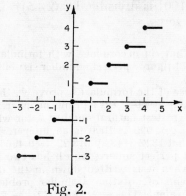

Fig. 2.

$$[\sqrt{7}] = 2, \quad \left[-\frac{19}{5}\right] = -4,$$
$$[6] = 6.$$

The function $[x]$ has "discontinuities", i. e. it varies by leaps. Figure 2 shows the graph of this function: the left end of each of the horizontal segments belongs to the graph (bold dots) and the right end does not belong to it.

Try to prove (7), that if $n! = p_1^\alpha p_2^\beta p_3^\gamma \ldots p$ (the canonical factorization of the number $n!$), then

$$\alpha = \left[\frac{n}{p_1}\right] + \left[\frac{n_2}{p_1}\right] + \left[\frac{n_3}{p_1}\right] + \ldots;$$ analogous formulae occur for $\beta, \gamma, \ldots, \sigma$.

Knowing this, it is easy to determine, for example, how many zeros are at the end of the number 100! Indeed, let $100! = 2^\alpha 3^\beta 5^\gamma \ldots 97^\sigma$. Then

$$\alpha = \left[\frac{100}{2}\right] + \left[\frac{100}{4}\right] + \left[\frac{100}{8}\right] + \left[\frac{100}{16}\right] + \left[\frac{100}{32}\right] + \left[\frac{100}{64}\right] +$$

$$+ \left[\frac{100}{128}\right] + \ldots = 50 + 25 + 12 + 6 + 3 + 1 = 97 \text{ and}$$

$$\gamma = \left[\frac{100}{5}\right] + \left[\frac{100}{25}\right] + \ldots = 20 + 4 = 24.$$

13

Therefore 100! is divisible by $(2+5)^{24}$ i. e. it ends in twenty four zeros.

1. Prove[8], that, in accordance with formula (6), when $\alpha =$ $= 2280$ a 1373-figure perfect number is obtained (1 log 2 $\simeq 0.301029996$).

2. Making use of the formula (3) prove [9], that for $N = 2a$ $(2^{a+1} - 1)$, where $2^{a+1} - 1$ is a prime number, $S(N) = 2\ N$.

3. Find the greatest natural number k, for which $101 \times 102 \times \ldots \times 999 \times 1000$ is an integer.[10]

4. Show [11] that 1322 314 049 613 223 140 496 $= 363\ 636\ 364^2$ is the smallest perfect square, which has two identical digits side by side when it is written down in the decimal system (in other systems of notation a similar problem has smaller solutions, for example $11_{(3)} = 2^2$; $882\ 882_{(33)} = 7332^2$; $288\ 288_{(33)} = 3900^2$). Seek out perfect cubes which are written down in some system of notation with two identical digits side by side (for example: $2^3 = 11_{(7)}$, 101101001 101101001$_{(3)} =$ $= 57$).

§ 3. Congruences

If the integers a and b, when divided by the natural number m, yield equal remainders, i.e. $a = mq_1 + r$ and $b = mq_2 + r$ (r, q_1, q_2 are integers, and $0 \leqslant r < m$), they are called *congruent, modulo m* and we write $a \equiv b$ (mod m). For example $27 \equiv -13$ (mod 8). Obviously the difference $a - b$ of two numbers a, b which are congruent modulo m is divisible by m. In our case, $27 - (-13) = 40$; 40 is divisible by 8.

We suggest, that the reader proves the following properties of congruences:

if $a \equiv b$ (mod m), $c \equiv d$ (mod m), then:
(1) $a + c \equiv b + d$ (mod m)
(2) $a - c \equiv b - d$ (mod m)
(3) $ka \equiv kb$ (mod m)   (k is any integer)
(4) $ac \equiv bd$ (mod m)
(5) $a^n \equiv b^n$ (mod m)   (n is any natural number)

(take into account, that $a \equiv b + mt$,
$c = d + mt'$ where t and t' are integers).

The enumerated properties easily yield ([12]) the theorem:

THEOREM. *If $\alpha \equiv \beta$ (mod m), and if $f(z) = a_0 + a_1 z + \ldots + a_n z^n$ is a polynomial with integral coefficients, then $f(\alpha) \equiv f(\beta)$ (mod m).*

This theorem helps in working out the tests for divisibility of a natural number N by 7, 9, 11, 13.

Let $N = c_k c_{k-1} c_{-k2} \ldots c_2 c_1 c_{0(10)} = c_k 10^k + c_{k-1} 10^{-1} + \ldots + c_2 10^2 + c_1 10 + c_0 = C_s 1000^s + C_{s-1} 1000^{s-1} + \ldots + C_2 1000^2 + C_1 1000 + C_0$; here $C_0, C_1, \ldots, C_{-1}$ are numbers, which are obtained, when the number N

is divided up from right to left into divisions of three digits each : $s = \left[\dfrac{k}{3}\right]$ and C_s may be a one-digit, two-digit or three-digit number (for instance, $N - 15\,032\,104\,341 = 341 + 104 \times 1000 + 32 \times 1000^2 + 15 \times 1000^3$; here $s = 3$, $C_0 = 341$, $C_1 = 104$, $C_2 = 32$ and $C_3 = 15$).

On introducing the notation

$$c_k z^k + c_{k-1} z^{k-1} + \ldots + c_2 z^2 + c_1 z + c_0 = f(z),$$
$$C_s z^s + C_{s-1} z^{s-1} + \ldots + C_1 z + C_0 = F(z),$$

we have

$N = f(10) = F(1000),$

$f(1) = c_k + c_{k-1} + \ldots + c_2 + c_1 + c_0 = \sigma(N)$ (the sum of digits of number N)

$f(-1) - c_0 - c_1 + c_2 - \ldots + (-1)^k c_k = \sigma'(N),$

$F(-1) - C_0 - C_1 + C_2 - \ldots + (-1)^s C_s = \sum'(N):$

we shall name the latter two sums, conditionally, *the algebraic sum of digits*, and *the algebraic sum of three-digit divisions* of the number N.

Since $10 \equiv 1 \pmod 9$, it follows from the last theorem that $f(10) \equiv f(1) \pmod 9$ or $N \equiv \sigma(N) \pmod 9$, i.e. N when divided by 9 gives the same remainder, as σN, therefore N is divisible by 9 when and only when σN is divisible by 9.

Similarly, from the congruence $10 \equiv -1 \pmod{11}$ we have: $f(10) \equiv f(-1) \pmod{11}$, or $N \equiv \sigma'(N) \pmod{11}$, therefore N is divisible by 11, if the algebraic sum of the digits of the number N is divisible by 11 (and conversely).

It follows from the easily verifiable congruences $1000 \equiv -1 \pmod 7$: $1000 \equiv -1 \pmod{11}$ and $1000 \equiv -1 \pmod{13}$ that

$F(1000) \equiv F(-1) \pmod 7, \quad F(1000) \equiv F(-1) \pmod{11}$

and

$$F(1000) \equiv F(-1) \pmod{13}.$$

or

$$N \equiv \sum{}'(N) \pmod 7, \quad N \equiv \sum{}'(N) \pmod{11},$$
$$N \equiv \sum{}'(N) \pmod{13},$$

i. e. N is divisible by 7 if the algebraic sum of the three-digit divisions of the number N is divisible by 7 (and conversely!); the tests for divisibility by 11 and 13 are formulated in the same way.

Similar arguments apply also in deducing the tests for divisibility of numbers, written down in the base-k system of notation, by $k-1$ and by $k+1$: the number N is divisible by $k-1$ (by $k+1$) when and only when the sum of its digits (the algebraic sum of digits), N being written down in the base-k system, is divisible by $k-1$ (by $k+1$). [Give the details of these arguments.]

Find the tests for divisibility by 5 and by 13 in the base-8 system, the tests for divisibility by 2, by 4 and by 7 in the base-3 system, the tests for divisibility by 13 and by 8 in the base-5 system of notation[13].

Congruences help to solve easily problems of the following type: find the remainder obtained in dividing the number $N = 13^{69} + 48 \times 10^{50}$ by 17. Obviously, we must find the smallest non-negative number congruent with N, modulo 17; applying the relevant property of congruences, we get: $13^{69} + 48 \times 10^{50} \equiv$
$\equiv (-4)^{69} - 3 \times 100^{25} \equiv 4 \times 16^{34} - 3 (-2)^{25} \equiv$
$\equiv -4 (-1)^{34} + 6 \times 16^{6} \equiv -4 \cdot 6(-1)^{6} \equiv 2 \pmod{17}$,
therefore, the required remainder equals 2.

Find[14] in the same way, the last two digits of numbers 293^{293}, 2^{1000}, $69^{69} + 31^{31}$.

Euler's Functions

The number of numbers, smaller than n (n is a natural number) and relatively prime to n, is called *Euler's Function* $\psi(n)$.

for $n=2$	3	4	5	6	7	8	9	10	12	20	36
$\varphi(n)=1$	2	2	4	2	6	4	6	4	4	8	12

(for example, $\psi(10) = 4$, since of the numbers, which are smaller than 10, only four numbers — 1, 3, 7, 9 are relatively prime to 10). We make the convention that $\psi(1) = 1$.

It is easy to prove([15]) that, when p is prime,

$$\psi(p) = p-1 \text{ and } \psi(p^k) = p^k - p^{k-1}.$$

In the theory of numbers, there is a theorem which states that for a and b which are relatively prime, $\psi(ab) = \psi(a)\psi(b)$. It follows hence, that, if $n = p_1^\alpha p_2^\beta \ldots p_k^\gamma$ is the canonical factorization of n, then

$$\varphi(n) = (p_1^\alpha - p_1^{\alpha-1})(p_2^\beta - p_2^{\beta-1}) \ldots (p_k^\lambda - p_k^{\gamma-1}). \qquad (1)$$

Gauss proved that the sum of the values of Euler's function, which have been calculated for all divisors of the number n, equals n. For instance

$$\varphi(1) + \varphi(2) + \varphi(5) + \varphi(10) = 1 + 1 + 4 + 4 - 10.$$

Euler proved that, for relatively prime k and n, it is always true that $k^{\varphi(n)} \equiv 1 \pmod{n}$: in particular, when p is prime and a is not divisible by p, $a^{p-1} \equiv 1 \pmod{p}$ ("Fermat's little theorem").

We recommend that the reader verifies the truth of Euler's, Gauss' and Fermat's theorems for a series of particular examples.

It follows from Euler's theorem, that the "exponential congruence" $k^z \equiv 1 \pmod{n}$, for relative primes k and n, is bound to have the solution $z - \psi(n)$: however, it may turn out that this congruence is true also for smaller values of z.

The smallest natural number z_0, satisfying this congruence is called *the index to which k belongs, modulo n*. It can be proved([16]) that z_0 must be a divisor of the number $\psi(n)$. In order to find, for instance, the smallest root of the congruence $60^z \equiv 1 \pmod{17}$, all divisors of the number $\psi(17) = 16$ must be tested. Since $60^z \equiv 9^z \pmod{17}$ and $9^2 \equiv -4 \pmod{17}$, $9^4 \equiv (-4)^2 \equiv -1 \pmod{17}$, $9^8 \equiv 1 \pmod{17}$, therefore 60 belongs to the index 8, modulo 17.

If the fraction $\frac{m}{n}$ in its smallest terms is represented in the form of a base-k fraction (k and n being relatively prime) the number of digits in the period of this fraction equals([17]) the index z_0, to which k belongs, modulo n.

Congruences

For instance, representing $\frac{1}{17}$ in the form of a systematic fraction in the 60-based system of notation, we shall have 8 digits in the period of the continued fraction. Make sure for yourselves by dividing 1 by 17, that $\frac{1}{17} = 0 \cdot (3 \ \overline{31} \ \overline{45} \ \overline{52} \ \overline{56} \ \overline{28} \ 7)_{(60)}$.

Find, in the manner indicated, the number of digits in the periods of the continued decimal fractions corresponding to the vulgar fractions $\frac{1}{7}$, $\frac{1}{13}$, $\frac{4}{13}$, $\frac{6}{13}$, $\frac{2}{19}$, and verify the results by direct division by 7, 13 and 19.

§ 4. Continued Fractions and Indeterminate Equations

Any positive rational number $\frac{a}{b}$ (a and b being natural numbers) can be represented in the form of a so-called (terminating) continued fraction.

Let the quotient obtained in dividing a by b be q_0 and the remainder be r_1; on dividing b by r, the quotient is q_1 and the remainder r_2; on dividing r_1 by r_2 the quotient is q_2 and the remainder r_3, etc. At some stage r_{n-1} must divide by r_n without a remainder $(r\frac{n-1}{r_n} = q_n)$.

In the theory of numbers it is proved that $r_n = (a, b)$ [(a, b) is the greatest common divisor of the numbers a and b]. The process of finding the greatest common divisor of two numbers by means of consecutive divisions is called *Euclid's algorithm*.

The division is usually laid out in the compact form:

$$
\begin{array}{r|l}
a & b \\
-\,bq_0 & \overline{q_0} \\
\hline
b & r_1 \\
-\,q_1r_1 & \overline{q_1} \\
\hline
r_1 & r_2 \\
-\,r_2q_2q_2 & \\
\hline
& r_3 \\
& \vdots \\
r_{n-1} & r_n \\
-\,q_nr_n & \overline{q_n} \\
\hline
0 &
\end{array}
$$

Obviously

$$\frac{a}{b} = q_0 + \frac{r_1}{b} = q_0 + \frac{1}{\dfrac{b}{r_1}} = q_0 + \cfrac{1}{q_1 + \dfrac{r_2}{r_1}} =$$

$$= q_0 + \cfrac{1}{q_1 + \cfrac{1}{\dfrac{r_1}{r_2}}} = \ldots = q_0 + \cfrac{1}{q_1 + \cfrac{1}{q_2 + \cfrac{1}{q_3 + \cfrac{}{\ddots}}}}$$

$$+ \cfrac{1}{q_{n-1} + \dfrac{1}{q_n}}.$$

This is the continued fraction required. It is often written down in the form:

$$\frac{a}{b} = q_0 + \cfrac{1}{q_1 + \cfrac{1}{q_2 + \cfrac{1}{q_3 + {}}}}$$

$$+ \cfrac{1}{q_{n-1} + \cfrac{1}{q_n}},$$

where any unity is divided by the whole expression underneath it.

It is also convenient to write this down in the agreed form

$$\frac{a}{b} = [q_0, q_1, q_2, \ldots, q_{n-1}, q_n],$$

where only the *partial denominators*

$$q_0, q_1, q_2, \ldots, q_{n-1}, q_n.$$

are shown.

For instance

$$\frac{173}{39} = 4 + \cfrac{1}{2 + \cfrac{1}{3 + \cfrac{1}{2 + \cfrac{1}{2}}}} = [4, 2, 3, 2, 2],$$

Because

$$
\begin{array}{r}
- \underline{173} \,\big|\, \underline{39} \\
156 \;\;\; 4 \\
\end{array}
$$

$$
\begin{array}{r}
-\underline{39} \,\big|\, \underline{17} \\
34 \;\;\; 2 \\
\end{array}
$$

$$
\begin{array}{r}
-\underline{17} \,\big|\, \underline{5} \\
15 \;\;\; 3 \\
\end{array}
$$

$$
\begin{array}{r}
-\underline{5} \,\big|\, \underline{2} \\
4 \;\;\; 2 \\
\end{array}
$$

$$
\begin{array}{r}
-\underline{2} \,\big|\, \underline{1} \\
2 \;\;\; 2 \\
\hline
0
\end{array}
$$

If the fraction is terminated at the k-th partial denominator, then, on representing the abbreviated continued fraction $q_0, q_1, \ldots, q_{k-1}, q_k$ in the form of a vulgar fraction, we obtain the so-called k-th convergent fraction $\frac{P_k}{Q_k}$. Obviously $\frac{P_n}{Q_n} = \frac{a}{b}$.

p Convergent fractions have a number of important roperties (see [2] or [28]).

Property 1. The numerators and denominators of three neighbouring terminating fractions are connected by a recurrence relationship

$$P_{k+1} = P_k q_{k+1} + P_{k-1}; \quad Q_{k+1} = Q_k q_{k+1} + Q_{k-1}, \tag{1}$$

enabling us to calculate easily the convergent fractions from known partial denominators.

Having worked out for the fraction $\frac{173}{39}$ $\frac{P_0}{Q_0} = 4 = \frac{4}{1}$

and $\frac{P_1}{Q_1} = 4 + \frac{1}{2} = \frac{9}{2}$, we can find the remaining convergent fractions by means of the formulae (1). The results of the calculation can be conveniently set out in a table

k	0	1	2	3	4
q_k	4	2	3	2	2
P_k	4	9	31	71	173
Q_k	1	2	7	16	39

For example;

$$P_2 = P_1 q_2 + P_0 = 9 \cdot 3 + 4 = 31,$$
$$Q_2 = Q_1 q_2 + Q_0 = 2 \cdot 3 + 1 = 7$$

and so on.

Property 2. Always

$$\frac{P_0}{Q_0} < \frac{P_2}{Q_2} < \frac{P_4}{Q_4} < \ldots < \frac{a}{b} = \frac{P_n}{Q_n} < \ldots < \frac{P_5}{Q_5} < \frac{P_3}{Q_3} < \frac{P_1}{Q_1}.$$

Property 3. For any k we have

$$\frac{P_k}{Q_k} - \frac{P_{k-1}}{Q_{k-1}} = \frac{(-1)^{k-1}}{Q_k Q_{k-1}} \qquad (2)$$

or

$$P_k Q_{k-1} - Q_k P_{k-1} = (-1)^{k-1}. \qquad (3)$$

It follows from here that $(P_k, Q_k) = 1$, because otherwise [i. e. when $(P_k, Q_k) > 1$] the expression $P_k Q_{k-1} - Q_k P_{k-1}$, which equals $(-1)^{k-1}$, is not divisible by the number (P_k, Q_k).

If $(a, b) = 1$, then it follows from $\frac{a}{b} = \frac{P_n}{Q_n}$, that $a = P_n$ and $b = Q_n$ and from (3) we have, when $k = n$.

$$a Q_{n-1} - b P_{n-1} = (-1)^{n-1}. \qquad (4)$$

The eqn. (4) is the key to the integral solution of the so-called *indeterminate* equation

$$ax + by = c, \qquad (5)$$

where a, b and c are integers, and $(a, b) = 1$.

Indeed, having rewritten (4) in the form

$$a(-1)^{n-1}cQ_{n-1}+b(-1)^{n}cP_{n-1}=c,$$

it is possible to state, that the numbers

$$x_0=(-1)^{n-1}cQ_{n-1}, \quad y_0=(-1)^{n}cP_{n-1} \qquad (6)$$

represent one of the integral solutions of the eqn. (5). It is easy to show[18] that:

(1) all remaining solutions of the equation (5) are obtainable from (6) by means of the formulae: $x = x_0 + bt$, $y = y_0 - at$, where t is an arbitrary integer;

(2) when $(a, b) > 1$, the equation $ax + by = c$ has no integral solution if c is not divisible by (a, b) (if, on the other hand c is divisible by (a, b), then, on dividing all terms of the eqn. (5) by (a, b) we arrive at the equation $a'x + b'y = c'$, where $(a', b') = 1$).

Let us clarify the above by means of an example.

In a jar, containing both spiders and beetles, there are altogether 38 legs. How many spiders (x) and how many beetles (y) are there in the jar, if a spider has 8 legs and a beetle 6?

Obviously, $8x + 6y = 38$, or $4x + 3y = 19$. Here $a = 4$, $b = 3$, $c = 19$. For the continued fraction $\frac{4}{3}=1+\frac{1}{3}$, $\frac{P_0}{Q_0}=\frac{1}{1}$ and $\frac{P_1}{Q_1}\equiv\frac{4}{3}$. Therefore

$$x_0=(-1)^{1-1}\times 19\times 1=19, \quad y_0=(-1)^1\times 19\times 1=-19,$$

whence

$$x=19+3t; \quad y=-19-4t.$$

The sense of the problem makes us interested in positive values of x and y only i. e. it should by $19 + 3t \geqslant 0$ and $-19 - 4t \geqslant 0$, or $-\frac{19}{3} \leqslant t \leqslant -\frac{19}{4}$. When $t = -5$ we have: $x_1 = 4$; $y_1 = 1$ (four spiders and one beetle) and when $t = -6$ we get $x_2 = 1$. $y_2 = 5$ (one spider and five beetles).

We shall indicate one more method of solving the eqn. (5). We rewrite it in the form $ax - c = -by$, Obviously, it is required to find integral values of x. for which the difference $ax - c$ is divisible by b, i. e.

$$ax \equiv c(\bmod b). \qquad (7)$$

This condition is satisfied by $x \equiv ca^{\varphi(b)-1}$ (mod b). Indeed, substituting this value in the congruence (7), we obtain $aca^{\varphi(b)-1} \equiv c \times 1$ (mod b) (the latter transition is made on the basis of Euler's Theorem).

For instance, for the equation $4x + 3y = 19$ we have $x \equiv 19 \times 4^{\varphi(3)-1}$ (mod 3) $\equiv 1 \times 1^{2-1}$ (mod 3) or $x = 1 + 3s$; but in that case $y = 5 - 4s$. For $s = 0$ we get $x_1 = 1$ and $y_1 = 5$ and for $s = 1$ we get $x_2 = 4$ and $y_2 = 1$.

Solve, by any of the methods shown, the indeterminate equations

$$\text{1) } 617x - 125y = 91, \quad \text{2) } 12x + 31y = 170$$

(see in [22] problems "auditing a cooperative" and "the trick of guessing someone's birth date").

It is also possible to expand any irrational number into a continued fraction. Separating out the integral part of α we get

$$\alpha = q_0 + \frac{1}{\alpha_1} \left(q_0 = [\alpha], \ \frac{1}{\alpha_1} < 1, \ \alpha_1 > 1 \right),$$

then

$$\alpha_1 = q_1 + \frac{1}{\alpha_2} \ (q_1 = [\alpha_1], \ \alpha_2 > 1),$$

$$\alpha_2 = q_2 + \frac{1}{\alpha_3} \ (q_2 = [\alpha_2], \ \alpha_3 > 1) \text{ etc.}$$

After repeating this operation n times we get $\alpha = [q_0, q_1, \ldots, q_{n-1}, \alpha_n]$: since for any n, α_n is irrational, the process never ends, and we obtain an *(infinite) continued fraction*;

$$\alpha = [q_0, q_1, q_2, \ldots, q_{n-1}, q_n, \ldots].$$

If α is a "quadratic irrationality" i. e. $\alpha = \frac{a + b\sqrt{c}}{d}$ where a, b, c, d are integers, then the partial denominators beginning with a certain number repeat themselves periodically (see [2]). For example:

$$\sqrt{3} = 1 + (\sqrt{3} - 1) = 1 + \cfrac{1}{\cfrac{1}{\sqrt{3} - 1}} = 1 + \cfrac{1}{\cfrac{\sqrt{3} + 1}{2}} =$$

$$= 1 + \cfrac{1}{1 + \cfrac{\sqrt{3} - 1}{2}} = 1 + \cfrac{1}{1 + \cfrac{1}{\cfrac{2}{\sqrt{3} - 1}}} =$$

$$= 1 + \cfrac{1}{1 + \cfrac{1}{\sqrt{3} + 1}} = 1 + \cfrac{1}{1} + \cfrac{1}{2 + (\sqrt{3} - 1)} .$$

Since $\alpha_3 = \sqrt{3} - 1 = \alpha_1$ therefore the partial denominators begin to repeat themselves in the future.

Commonly, we should write; $\sqrt{3} = [1, 1, 2, 1, 2, 1, 2 \ldots]$.

Show by a similar method that $\sqrt{5} = [2, 4, 4, \ldots]$ and $\sqrt{7} = [2, 1, 1, 1, 4, 1, 1, 1, 4, 1, 1, \ldots]$.

The properties of terminating fractions shown above, which also hold for infinite continued fractions, permit the easy finding of rational numbers as close as desired to α, provided sufficient partial denominators are known.

Indeed, on the basis of the second property of convergent fractions, α is enclosed between any neighbouring convergent fractions $\dfrac{P_{k-1}}{Q_{k-1}}$ and $\dfrac{P_k}{Q_k}$ But the absolute value of their difference equals $\dfrac{1}{Q_{k-1}Q_k}$ (third property). Therefore the error of the approximate equation $a \approx \dfrac{P_{k-1}}{Q_{k-1}}$ is less than $\dfrac{1}{Q_{k-1}Q_k}$ For example, for $\sqrt{5} = [2, 4, 4, 4 \ldots]$ we have

k	0	1	2	3	4 ...
q_k	2	4	4	4	4 ...
P_k	2	9	38	161	682 ...
Q_k	1	4	17	72	305 ...

26

therefore, $\sqrt{5} \simeq \frac{38}{17} \left(\text{error} < \frac{1}{17 < 72} \right)$, $\sqrt{5} \simeq \frac{161}{72}$ (error$<$

$\frac{1}{72+305}$), etc.

Find (^{19}) convergent fractions for $\sqrt{2} = [1, 2, 2, \ldots]$ and for $\sqrt{3} = [1, 1, 2, 1, 2, \ldots]$, which differ from $\sqrt{2}$ and from $\sqrt{3}$ respectively by less than 10^{-6}.

The expansion of $\sqrt{m}$ into a continued fraction gives a simple method of solving in whole numbers the so-called equation of Pell

$$x^2 - my^2 = 1. \tag{8}$$

Let $\sqrt{m} = [q_0, q_1, q_2, \ldots, q_{s-1}, \underline{q_s, q_1, q_2, \ldots}]$ (the period of the continued fraction is underlined).

It turns out (see [2]) that, for s even, the equation (8) has as its solutions the pairs of numbers (P_{s-1}, Q_{s-1}), (P_{2s-1}, Q_{2s-1}), (P_{3s-1}, Q_{3s-1}) etc., and for s odd, the pairs of numbers (P_{2s-1}, Q_{2s-1}), (P_{4s-1}, Q_{4s-1}), (P_{6s-1}, Q_{6s-1}) etc.

In [2] a table is given of expansions of $\sqrt{m}$ into a continued fraction, when $m < 100$, and also the smallest whole positive solutions of the equations $x^2 - my^2 = 1$.

For $\sqrt{10} = [3, 6, 6, 6, \ldots]$ we have

k	0	1	2	3 ...
q_k	3	6	6	6 ...
P_k	3	19	117	721 ...
Q_k	1	6	37	228 ...

Since $s = 1$, the solutions of the equation $x^2 - 10y^2 = 1$ are the pairs of numbers $x_1 = P_1 = 19$ and $y_1 = Q_1 = 6$; $x_2 = P_3 = 721$ and $y_2 = Q_3 = 228$, etc.

For $\sqrt{32} = (5, 1, 1, 1, 10, 1, 1, \ldots]$ we have

k	0	1	2	3	4	5	6	7
	5	1	1	1	10	1	1	1
P_k	5	6	11	17	181	198	379	577
Q_k	1	1	2	3	32	35	67	102

Since $s = 4$, the solutions of the equation $x^2 - 32y^2 = 1$ are the pairs of numbers $x_1 = P_3 = 17$ and $y_1 = Q_3 = 3$; $x_2 = P_7 = 577$ and $y_2 = Q_7 = 102$ etc.

Given that $\sqrt{89} = [9, \overline{2, 3, 3, 18}, 2, 3, \ldots]$ and $\sqrt{61} = [7, \overline{1, 4, 3, 1, 2, 2, 1, 3, 4, 1, 14}, 1, 4, 3, \ldots]$, show that the smallest integral solutions of the equations $x^2 - 89y^2 = 1$ and $x^2 - 61y^2 = 1$ are the pairs of numbers (500 001, 53000) and 1766 319 049, 226 153 980) respectively.

In an article by the Polish mathematician Sierpinski the smallest solution of the equation $x^2 - 991y^2 = 1$ is given: $x_1 = 379\ 516\ 400\ 906\ 811\ 930\ 638\ 014\ 896\ 080$ and $y_1 = 12\ 055\ 735\ 790\ 331\ 359\ 447\ 442\ 538\ 767$.

This means that for any integral value of y not exceeding the number y_1, $\sqrt{(991y^2 + 1)}$, is an irrational number, and the rational number x_1 is obtained, when at least $y = y_1$.

In order to imagine how great are the numbers x_1 and y_1 in the last example, note, that if we were to take a milliard (10^{18}) years to work out values of $\sqrt{(991y^2 + 1)}$ for $y = 1, 2, 3, 4, \ldots$ using one second for each calculation, we should not achieve an exact extraction of the square root.

Yet we could not maintain that $\sqrt{(991y^2 + 1)}$ is irrational for any natural y; once we increase the "testing time" approximately 400 times, we arrive at y, and we discover that $\sqrt{(991y^2 + 1)}$ is a rational number.

Let us consider one more problem, ascribed to Archimedes, and reducible in the end to an equation of type [8].

Continued Fractions and Indeterminate Equations

In a manuscript discovered at the end of the eighteenth century it is said that Archimedes had found an ancient inscription and had sent the problem contained in it to the mathematicians of Alexandria.

The problem was to determine the size of the herd of cows and oxen belonging to the Sun. It followed from the first part of the data, given in verse in the problem (see [16] pp. 204–206), that the herd consisted of white, black, brown and dappled oxen and cows, and the numbers of oxen (U, X, Y, Z) and cows (u, x, y, z) of the different hues were connected by the relationships

$$U = \frac{5}{6} X + Y; \quad X = \frac{9}{20} Z + Y; \quad Z = \frac{13}{42} U + Y;$$

$$u = \frac{7}{12} (X + x); \quad x = \frac{9}{20} (Z + z); \quad z = \frac{11}{30} (Y + y);$$

$$y = \frac{13}{42} (U + u).$$

Amateur calculators can find ([20])

U = 10 366 482 t,	u = 7 206 360 t,
X = 7 460 514 t,	x = 4 893 246 t,
Y = 7 358 060 t,	y = 3 515 820 t,
Z = 4 149 387 t,	z = 5 439 123 t

(t takes any integral values).

However, in the second part of the text of the problem, which is directed to the person who is to solve it, it is noted that the chief difficulties of the problem are those created by the supplementary conditions;

"If you count up how many cattle there were,
How many fat oxen and
How many milch-cows of various hues there were
No one dare call you a fool at numbers.
Yet do not call yourself wise if you fail
To reckon with different ways of the ox."

After enumerating the different ways and habits of oxen, the concluding part of the text of the problem says:

"Find all this and with the eye of your soul
encircle the herd and be able to pass
on your knowledge. Then step forth in pride;
Victory is yours and you're wisest of all."

If the habits of oxen are taken into account, then t
in the given formulae has to be selected in such a way
that

(1) the sum $U + X$, which equals $17826\,996t$, is a
perfect square, for which t should be taken as
equal to $4\,456\,749s^2$, where s is any natural
number;

(2) the sum $Y + Z$, which equals $11\,507\,447t$, is a
"triangular number", i. e. a number of the form
$\frac{n(n+1)}{2}$.

By substituting the expression obtained above in
place of t, we come to the equation $51\,285\,803\,909\,803\,s^2$
$= \frac{n(n+1)}{2}$.

If we multiply both sides of this equation by 8, then
add 1 to both sides, and, finally substitute w for $2n + 1$,
we obtain the Pellian equation;

$$w^2 - 410\,286\,423\,278\,424\,s^2 = 1.$$

If we denote the number of cattle in the herd by N
(its smallest value), then when all the supplementary
conditions are taken into account, it turns out (according to calculations published in 1880 by Amthor) that

$$N \simeq 77 \times 10^{206\,543}.$$

It is doubtful whether the reader will manage "with
the eye of his soul to encircle the herd and be able to
pass on his knowledge"!

§ 5. Pythagorean and Heronic Triples

The well-known relationship between the hypotenuse and the other two sides of a right-angled triangle, $x^2 + y^2 = z^2$, can be considered as an indeterminate equation with three unknowns.

It turns out [2], that all possible triples of integers which are relatively prime in pairs, and that satisfy the above equation ("Pythagorean triples"), are obtainable from the formulae;

$$\begin{cases} x = u^2 - v^2 \\ y = 2uv \\ z = u^2 + v^2 \end{cases}$$

if we give the auxiliary variables relatively prime values and make one of them even, while the other one is odd. (If this condition is not observed, we obtain Pythagorean Triples, whose highest common factor is greater than 1.)

For example

u	v	x	y	z
2	1	3	4	5
4	1	15	8	17
3	2	5	12	13
4	3	7	24	25
5	2	21	20	29
3	1	8	6	10

Pythagorean Triples are just a particular case of "Heronic Triples" — the name for three integers expressing the lengths of the sides of a triangle with an integral area.

31

It is easy to prove([21]) that any of the altitudes of a "Heronic triangle" (for instance, BD in Fig. 3) gives two right-angled triangles (ABD and BDC) with rational sides, either adjacent or overlapping each other.

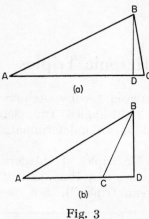

(a)

(b)

Fig. 3

If we take triangles $\triangle$ and $\triangle'$ with integral sides a, b, c and a', b, c' (c and c' are the hypotenuses) and we multiply all sides of $\triangle$ by $\frac{a'}{a}$ we obtain the triangle $\triangle$, with sides a', $\frac{ba'}{a}$, $\frac{ca'}{a}$, similar to the triangle $\triangle$.

By causing the identical sides of the triangles $\triangle$, and $\triangle'$ to become common to them both, we obtain two triangles with rational sides $\frac{a'}{a}c$, c', $\left|\frac{a'}{a}b \pm b'\right|$. Multiplying these numbers by a, we obtain two Heronic triples $a'c$, ac', $|a'b \pm b'a|$.

In the same way, equalizing the side a with the side b' (multiplying the sides of triangle $\triangle$ by $\frac{b'}{a}$), or equalizing the side b with the side a' [multiplying by $\frac{a'}{b}$, or equalizing b with b' (multiplying by $\frac{b'}{b}$)] we can get six more Heronic triples:

$$a'c, bc', |a'a \pm b'b|: \quad b'c, ac', |b'b \pm a'a|:$$
$$b'c, bc', |b'a \pm a'b|.$$

For instance, for the Pythagorean triples (3, 4, 5) and 15, 8, 17) it is easy to find, by the method shown, 8 Heronic triples; (75, 51, 84), (75, 51, 36), (75, 68, 77), (75, 68, 13), (30, 51, 77), (40, 51, 13), (40, 68, 84), (40, 68, 36). The first two and the last two triples, when divided by 3 and 4 respectively, give: (25, 17, 28), (25, 17, 12), (10, 17, 21), (10, 17, 9).

Find ([12]) Heronic triples using the Pythagorean triples
  (1) (3, 4, 5) and (5, 12, 13),
  (2) (7, 24, 25) and (7, 24, 25).

§ 6. Arithmetical Pastimes

There are arithmetical problems, whose solution is not connected with any theory, and require of the solver only ingenuity and patience.

Problems of this kind include the search for interesting relationships between numbers, for numerical curiosities, and so on.

Let us quote several typical examples;

1. Distribute symbols of arithmetical operations and, if necessary, brackets between the numbers 1 2 3 4 5 6 7 8 9, without changing their order, in such a way that the result should be a given number N. For example, when $N = \frac{1}{2}$ and when $N = 1$, we have: $\frac{1}{2} = (123-45):(67+89)$; $1 = 1 + 2 - 3 + 4 - 5 - 6 + 7 - 8 + 9$; $1 = 1 + 23 - 45 - 67 + 89$ and so on.

One may set oneself the task of representing in this way the greatest possible number of natural numbers or fractions (for instance, fractions of the form $\frac{k}{5}$, where $k = 1, 2, 3, \ldots$). Or conversely, one could take some number and try to represent it in all possible ways.

Certain restrictions may be introduced into the problem, for example by permitting the use of the signs $+$ and $-$ only: or conversely, freedom of action may be increased by permitting the use of radicals or changing the order of the digits (for instance, $100 = 67^2 - 4385 - 1 - \sqrt{9}$) etc.

2. Distribute the digits 1, 2, 3, 4, 5, 6, 7, 8, 9 in given schemes in such a way that on carrying out the operations indicated correct equalities are obtained

33

(a) $- - \times - - - - = - - - -$ (e.g. $12 \times 483 =$
$= 5796$, etc.)

(b) $- - - \times - - - = - - \times - - -$

(c) $- - - - \times - - = - - \times - - -$

3. Distribute the digits 1, 2, 3, 4, 5, 6, 7, 8, 9 in such a way, that the product of three three-digit numbers: $- - - \times - - - - \times - - - -$ is as great as possible (or as small as possible).

4. Distribute the digits 0, 1, 2, 3, 4, 5, 6, 7 8, 9 in such a way that $- - - - - - \div - - - - - - = n$, where n equals one of the numbers 2, 3, 4, 5, 6, 7, 8, 9 (see [39], 1945, No. 3).

5. Certain numbers can be represented in a different form, without introducing new digits: e. g. $660 =$ $= 6! - 60$; $1395 = 15 \times 93$; $145 = 1! + 4! + 5!$; $144 = (1 + 4)! + 4! = (1 + \sqrt{4})! \times 4!$ $387\,420\,489 =$ $= 3^{87 + 420 - 489}$.

Try to find similar equalities.

6. By using once each of the digits 1, 2, 3, 4, 5, 6 and each of the operations — addition, subtraction, multiplication, division and raising to a power — obtain the greatest possible number ([36], 1946, No. 2, p. 49).

7. In the expression $1 \div 2 \div 3 \div 4 \div 5 \div 6 \div 7 \div 8 \div 9$ brackets should be distributed in such a way, that after calculations, the greatest possible (or the smallest possible) number is obtained.

8. In the examples $41096 \times 83 = 3410968$ and $8 \times \times 86 = 688$ the multiplication by a two-digit number has been simplified; the second digit of the multiplier is placed at the beginning of the multiplicand and the first at the end.

Look for similar examples.

9. It is mentioned in a certain journal that within the bounds of the first ten thousand there exist only eight numbers which can be written down in two systems of notation by three identical digits.

The smallest and the largest of them are $273 = 111_{16} = 6$ $= 333_{(9)}$ and $9114 = 222_{(67)} = \overline{14}\ \overline{14}\ \overline{14}_{(25)}$; try to find the rest.

Could you find the numbers (*a*) written down in three systems of notation by three identical digits; (*b*) written down in two systems by four identical digits.

10. Enthusiasts can be recommended to collect existing numerical curiosities and seek new ones. We quote here several examples which can suggest themes for independent researches.

a) $\overline{10\ 10}\ 4\ 4_{(13)} = \overline{11\ 11}^2_{(13)}$; $4\ 4\ \overline{10\ 10}_{(13)} = 7\ 7^2_{(13)}$

b) $7778^2 - 2223^2 = 55\ 555\ 555$;

c) $888\ 889^2 - 111\ 112^2 = 777\ 777\ 777\ 777$;

d) $999\ 999\ 999^2 = 12\ 345\ 678\ 987\ 654\ 321 \times (1+2+ \\ +3+4+5+6+7+8+9+8+7+6+5+4+3+2+1)$.

How would you verify the truth of the last three equations in the simplest possible way?[23]

11. *Amusing cancelling.* There are fractions whose magnitude does not alter on crossing out identical digits, or even groups of digits, in the numerator and the denominator. For instance; $\dfrac{19}{95} = \dfrac{1}{5}$; $\dfrac{3544}{7531} = \dfrac{344}{731}$; $\dfrac{2666}{6665} = \dfrac{266}{665} = \dfrac{26}{65} = \dfrac{2}{5}$; $\dfrac{143185}{17018560} = \dfrac{1435}{170560}$; $\dfrac{4251935345}{91819355185} = \dfrac{425345}{9185185}$ etc. It is possible to set oneself the task of finding all fractions admitting such "cancelling" by discussing the number of digits in the numerator and the denominator and what places should the "cancellable" digits occupy. For instance, it is easy to obtain from $\dfrac{10a+b}{10b+c} = \dfrac{a}{c}$ ($a \neq c$), that $c = \dfrac{10ab}{9a+b}$; within the bounds of ten, integral values of *b* and *c* are obtained only for $a = 1, 2, 4$, which gives fractions $\dfrac{16}{64}$, $\dfrac{19}{95}$, $\dfrac{26}{64}$, $\dfrac{49}{98}$.

12. Using four 4's (other variants require five 5's, four 5's etc.) represent any number from 1 to *n* where *n* is a natural number. For example; $1 = 4 - 4 + \dfrac{4}{4}$; $7 = \dfrac{44}{4} - 4$ etc.

We can agree to use the "factorial" sign $\left(21 = 4!\right.$ $+ \frac{4}{4} - 4\left.\right)$ the "radical" sign $\left(17 = \sqrt{4^4} + \frac{4}{4}\right)$, and to introduce a "point above the figure" to represent recurrent decimal fractions. For example: $19 = \frac{4}{.4} + \frac{4}{.4\dot{}}$ and so on.

If we admit the use of the logarithm sign, then, as follows from the formula

$$n = -\log_4\left(\log_4 \underbrace{\sqrt{\sqrt{\sqrt{\ldots\sqrt{\sqrt{4}}}}}}_{2\pi \text{ radicals}}\right) \quad \text{check!}$$

it is even possible to express any natural number n by means of three 4's.

§ 7. Numerical Tricks

One sometimes comes across people who carry out mentally and with phenomenal speed, operations with multi-digit numbers. Such calculators, by utilizing a number of numerical tricks and various artificial steps, astonish audiences with one or two spectacular acts.

Of course, even when knowing the secret of this or that "act", not everybody can be successful as a performer. The knowledge of the secret helps in the simplest cases only.

Let us describe several simple numerical tricks.

Guessing a Number, Thought of by Someone

Having asked someone to think of a number and to carry out certain definite operations with it, it is easy to determine the number thought of from the result of the calculations.

Here are several examples:

1. Ask someone to think of a number, to add 3 to it, multiply the sum by 6, subtract the number thought of from the product, subtract another 8, and, finally, divide the rest by 5. When you are given the result, you can state the number thought of straight away. As it follows from the equation $[(x + 3) \times 6 - x - 8] \div 5 = x + 2$, it is sufficient to diminish the result obtained by 2.

2. Ask someone to think of two numbers x and y, up to one hundred (the day of the month, the age of a

person, the size of shoes, the change of a rouble, etc.) and carry out a series of operations which are determined by the left-hand side of the equation

$$(2 \times x + 5) \times 50 + y - 365 = 100x + y - 115 = N.$$

Then, knowing N, you can determine x and y (having added 115 to N and split N into two numbers by sepaating the two last digits).

3. Ask someone to think of the day (l) of the month (m) of the year (n) of some event that happened in the twentieth century (the last two figures of the year should be taken) and carry out operations indicated in the left-hand side of the equation

$$[(20l + 222) \times 5 + m] \times 100 + n + 111 =$$
$$= 10\ 000l + 100m + n + 111\ 111 = N.$$

on l, n and m. In order to determine the date thought of, 111 111 must be subtracted from N and the difference split up into groups of two digits each, from right to left. For example, if $N = 201656$, the date thought of was 9. 05. 45, i. e. 9 May 1945. Any number of similar tricks can be invented.

The Guessing of the Results of Operations with an Unknown Number

There exist a number of tricks based on the fact that certain definite operations give identical results for a fairly wide class of numbers. This sometimes occurs due to the exclusion of the number thought of in the process of carrying out the operations, at other times because of the properties of the class of numbers or of the operations which are carried out.

1. If you write down on the right- hand side of any three-digit number N the same number once again, and divide the six-digit number thus obtained (it equals $1001 \times N = N \times 7 \times 11 \times 13$) by 7, divide the quotient obtained by N, finally, divide the second quotient by 11, the number resulting is always 13.

38

Most people find it surprising that all divisions come out exactly, although N is chosen at random.

2. Since any odd prime number other than 3 can be represented in the form $6k \pm 1$, therefore $p^2 + 17 = = 36k^2 \pm 12k + 18$, i. e. $p^2 + 17$ always yields the remainder 6 on being divided by 12.

3. If in a three-figure number $\overline{abc}$ (a, b, c are the digits of the number) $a > c$, then:

(1) in the difference $\overline{abc} - \overline{cba} = \overline{\alpha\beta\gamma}$, $\beta = 9 = \alpha + \beta$ always,

(2) $\overline{\alpha\beta\gamma} + \overline{\gamma\beta\alpha} = 1089$.

Thus, knowing only one of the end digits of $\overline{\alpha\beta\gamma}$ it is possible to state at once the difference between the number $\overline{abc}$ which was thought of and the inverted number, and, knowing nothing of $\overline{\alpha\beta\gamma}$ it is possible to guess the outcome of adding $\overline{\alpha\beta\gamma} + \overline{\gamma\beta\alpha}$ for any number thought of.

The Determination of a Number Thought of, Using Three Tables

When numbers from 1 to 60 are distributed in their order in each of three tables, in such a way, that the first table is made up of three columns of twenty numbers each, the second one of four columns with 15 nubers in each, and the third of five columns of 12 numbers each (see Fig. 4), it is easy to determine quickly the number N ($N \leqslant 60$) thought of by someone, if that person points out the numbers, α, β, γ, of the columns in which the number thought of occurs in each of the three tables; N equals the remainder of the division of the number $40\alpha + 45\beta + 36\gamma$ by 60 or, in other words, N equals the smallest positive number, congruent with the sum $(40\alpha + 45\beta + 36\gamma)$, modulo 60. For example, for $\alpha = 3$, $\beta = 2$, $\gamma = 1$, $40\alpha + 45\beta + 36\gamma \equiv \equiv 0 + 30 + 36 \equiv 6 \pmod{60}$ i. e. $N = 6$ [24].

A similar question can be solved for numbers up to 420, distributed in four tables with three, four, five and seven columns: if α, β, γ, δ are the numbers of columns in which the number thought of is to be found then it equals the remainder of the division of the

I	II	III
1	2	3
4	5	6
7	8	9
•	•	•
•	•	•
•	•	•
55	56	57
58	59	60

I	II	III	IV
1	2	3	4
5	6	7	8
•	•	•	•
•	•	•	•
•	•	•	•
53	54	55	56
57	58	59	60

I	II	III	IV	V
1	2	3	4	5
6	7	8	9	10
•	•	•	•	•
•	•	•	•	•
51	52	53	54	55
56	57	58	59	60

Fig. 4

numbers $280\alpha + 105\beta + 336\gamma + 120\delta$ by 420. Attempt to prove this, extending somewhat the arguments put forward in § 38 (see note[24]).

A Card Trick

Having asked someone to remove one card from a pack of 36 cards, it is possible to guess the name of the card by glancing quickly at each remaining card in turn (while holding the pack in one's hand). Here it is unnecessary to attempt to retain in one's memory the cards that remained in the pack, but it is sufficient to calculate the sum S of the remaining points, which is quite simple, given some practice. Counting an ace as one point for simplicity, we have for $S : 190 \leqslant S < 200$ and therefore it is only necessary to find the number of units in S (discarding the tens).

If that number equals, for example, 3, then $S = 193$, and it follows that the card removed from the pack was a seven, whose suit can be easily established on a second quick glance through the pack.

Who Took What?

The person who is to do the guessing gives one, two and three coins to three persons, A, B and C, respectively and leaves 18 more coins on the table. *A, B* and *C*, in his absence, distribute among themselves three objects: a fork, a spoon and a knife, after which the possessor of the fork takes as many additional coins as he received in the first place, and the possessors of the spoon and the knife respectively take double and four times as many coins, as they received originally.

For each of the six possible distributions of objects among *A, B* and *C*: *f, s, k; s, f, k; f, k, s; s, k, f; k, f, s: k, s, f*, there will be 1, 2, 3, 4, 5, 6, 7 coins left on the table, respectively. If you invent a simple method for remembering the variants of distribution of three objects among three persons in the same sequence as that written down above, you will be able successfully to guess who took what, from the number of coins left on the table.

In order to determine the owners of *n* different objects in a similar way — by coins left — it is necessary to select *n* different numbers $a_1, a_2, \ldots a_n$ (for the initial distribution of coins) and *n* multipliers, $m_1, m_2, \ldots, m_n$ such that for different distributions of these multipliers among the numbers $a_1, a_2, \ldots, a_n$ there would arise *n*! unequal, but differing as little as possible, sums of the form

$$m_{a_1}a_1 + m_{a_2}a_2 + \ldots + m_{a_n}a_n$$

($\alpha_1, \alpha_2, \ldots, \alpha_n$ are the number-names of objects, taken by the 1st, 2nd ..., *n*th person, respectively). For *n* = 4 it is possible to take, for example $a_1 = 1, a_2 = 2, a_3 = 3, a_4 = 4$ and the multipliers m_1:1, 2, 5, 15 (see [30] Ch. 12).

The Extraction of Roots of Multidigit Numbers

By making use of a number of simple procedures, it is fairly easy to extract mentally odd roots from numbers which are exact powers of two-digit or even three-digit numbers.

When working out $\sqrt[s]{N}$, the number N should be split up into divisions, from right to left, each division containing s digits (the last division on the left may contain fewer digits, of course).

The last digit of the root can be easily determined by using the following two rules:

I. Numbers n^5, n^9, n^{13}, ..., n^{4S+1} always end in the same digit as n.

II. If the last digit of n is 0, 1, 2, 3, 4, 5, 6, 7, 8, 9, then n^3, n^7, n^{11} ..., n^{4S-1} end with 0, 1, 8, 7, 4, 5, 6, 3, 2, 9 respectively.

The first digit of the root $\sqrt[s]{N}$ when $s = 3$ is obtained easily if the values n^3 are known for $1 \leq n \leq 9$; when $s = 5, 7, 9$, etc., the table of logarithms of the first nine numbers (to the second decimal place) should be memorized;

n	1	2	3	4	5	6	7	8	9
$\log n$	0	0·30	0·48	0·60	0·70	0·78	0·85	0·90	0·95

E x a m p l e s. 1. It is possible to write down immediately $\sqrt[3]{314432} = 68$, by applying rule II and noting that $6^3 < 314 < 7^3$.

2. $\sqrt[7]{N} = \sqrt[7]{17565568854912} = n$. Since $10^{13} < N < 2 \times 10^{13}$,
therefore $13·0 < \log N < 13·3$,
(consequently)$1·85 < \log n = \frac{1}{7} \log N < 1·9$, i. e., in accordance with the table shown, $70 < n < 80$: on applying rule II we get $n = 78$.

3. The 25th root of a 48-digit number ending in 8 equals 68 ($45 < \log N < 46$, consequently $1·8 < \log n < \frac{1}{25} \log N < 1·84$).

If N ends in one of the digits 1, 3, 7, 9, it is possible to find the last two digits of $\sqrt[3]{N}$ if we know the last

two digits of the number N. For instance; $\sqrt[3]{\ldots 53} =$
$= n = 10z + 7$. We work out mentally: $(7 + 10z)^3 =$
$= 343 + 147 \times z \times 10 + \ldots$ (the last two terms do
not influence the number of units or the number of
tens).

Since N has five tens and 343 has four tens, the number $147z$ ends in 1 and z ends in 3 i. e. $n = 100y + 37$.

This procedure sometimes enables calculators, having asked someone to dictate slowly the digits of N from right to left, to give the answer (for $N < 10^6$) before the end of dictation.

When certain other circumstances are taken into account, this procedure can be used even when the last digit of N differs from 1, 3, 7 or 9; it can be applied, when $s = 5$, 7, etc. While referring those who are interested to [18], where several procedures used by performing calculators are described, we shall give here the description of one more method of finding one of the digits of the number $n = \sqrt[3]{N}$, when knowing the remaining digits of n.

This method is based on the following rule([25]):
if, on dividing n^3 by 11, we obtain the remainder d:
0, 1, 2, 3, 4, 5, 6, 7, 8, 9, 10
then, on dividing n by 11, we get the remainder d:
0, 1, 7, 9, 5, 3, 8, 6, 2, 4, 10.

As it is known from § 3, $N \equiv \sigma'(N)$ (mod 11) where $\sigma'(N)$ is the algebraic sum of digits of the number N. Therefore it is necessary to find $\sigma'(N)$ and determine d, which is congruent with it, modulo 11 $(0 \leq d \leq 10)$.

If the correspondence between d and d_1 is remembered (and lightning calculators have to keep plenty of things in mind!) it only remains to find the one unknown digit of the number n required from the known d_1 and the condition $d_1 \equiv \sigma'(n)$ (mod 11). Example:
$\sqrt[3]{54\,053\,028\,541} = 3z\,81 = n$. (How to determine all the other digits, apart from Z, is described above.)

As $\sigma'(N) \equiv 1 - 4 + 5 - 8 + 2 - 0 + 3 - 5 + 0 - 4 + 5 = -5$ and $-5 \equiv 6$ (mod 11), therefore

$d = 6$, consequently $\sigma'(n) = 1 - 8 + Z - 3 = Z - 10$ should be congruent with $d_1 = 8$, modulo 11, whence $z = 7$.

It goes without saying, that only a person with great aptitude for calculations, and having been specially trained, would be capable of performing all these calculations mentally at a speed.

§ 8. Rapid Calculations

The desire to simplify operations with many-figure numbers brought about the invention, in the 17th century, of tables of logarithms, and thereafter of the logarithmic ruler.

In the 19th century arithmometers made their appearance, and the beginning of the 20th century saw the advent of automatic calculating machines; about fifteen years ago there appeared electronic computers which are capable of solving problems, requiring millions of operations with multidigit numbers, within a matter of hours.

Abbreviated Multiplication

In multiplying, say, the numbers 7496 and 3852 we shall write down the complete products of each digit of the multiplicand by each digit of the multiplier in their respective orders.

			× 7	4	9	6	
			3	8	5	2	
0	0	0	14	8	18	12
0	0	35	20	45	30	
0	56	32	72	48		
21	12	27	18			

2	8	8	7	4	5	9	2
	$\widehat{7}$	$\widehat{10}$	$\widehat{13}$	$\widehat{10}$	$\widehat{4}$	$\widehat{1}$	

45

Any vertical column enclosed by the horizontal lines contains (from the top) (1) the product of the unit digit of the multiplier and the corresponding digit (α) of the multiplicand, (2) the products of the remaining digits of the multiplier, taken from right to left and the digits of the multiplicand to the right of α, taken simultaneously from left to right.

For example, in the hundreds we have:

$$8(=2\times4), \quad 45(=5\times9), \quad 48(=8\times6);$$

here α = 4. In the ten thousands we have:

$$0(=2\times0), \quad 35(=5\times7), \quad 32(=8\times4), \quad 27(=3\times9);$$

here α = 0.

Adding up the numbers in each column and adding to them the "tens" carried in memory, which were obtained in adding up the numbers in the neighbouring order on the right (see numbers under the little arcs) we obtain, from right to left, all the numbers of the required product.

The numbers in the rows joined by the figured bracket, do not have to be set down in writing, and addition of the corresponding two-digit numbers can be carried out mentally. For this purpose, it is necessary, of course, to learn to add up two-digit numbers faultlessly and at a high speed.

For example having arrived at the order of hundreds (we have carried 4 from the order of tens) we announce (looking at 2 and 4 and then at 5 and 9, and finally at 8 and 6); "8 and 4 is 12, and 45 is 57, and 48 is 105, carry 10" (and we write 5 in the order of hundreds).

Practise first with two-digit and then with three-digit, etc. numbers, and you will see that after a while it is possible to multiply multidigit numbers easily and quickly.

Abbreviated Division

In dividing multidigit numbers it is easy to learn, while multiplying the divisor by the current digit of the quotient, to subtract simultaneously (without writing it down) the product thus obtained from the number N, which is formed by the first few digits of the dividend, or by one of the intermediate remainders. The digits of the difference required are determined in such a way (from right to left) as to give the number N, when added to the above product.

For instance: $\frac{496343}{328} \bigg| \frac{927}{5}$. Here the digits of the difference 8, 2, 3 are obtained as follows: noting mentally that $5 \times 7 = 35$ and that 8 has to be added in order to obtain a number ending in 3, we say "35 and 8 is 43, carry 4 (we write down the digit 8 of the difference); 4 and 10 (5×2) is 14, and 2 is 16, carry 1 (we write down the digit 2 of the difference), 1 and 45 (5×9) is and 3 is 49" (we write down the digit 3 of the difference required.

The Abacus and Napier's Rods

The multiplication and division of multidigit numbers is considerably simplified, when an abacus and so-called Napier's rods are available.

Each one of the Napier's rods gives products of the digits in its "heading" and 1, 2, 3, 4, 5, 6, 7, 8, 9 (see Fig. 5a). Here, the tens appear above the sloping line in the corresponding box, and the units below it.

When several rods are laid side by side in such a way that their headings form the multiplicand, and if an auxiliary rod is laid alongside on the left, to indicate the numbers of rows, the product of the multiplicand in the heading and 1, 2, 3, 4, 5, 6, 7, 8, 9 are easily read off directly.

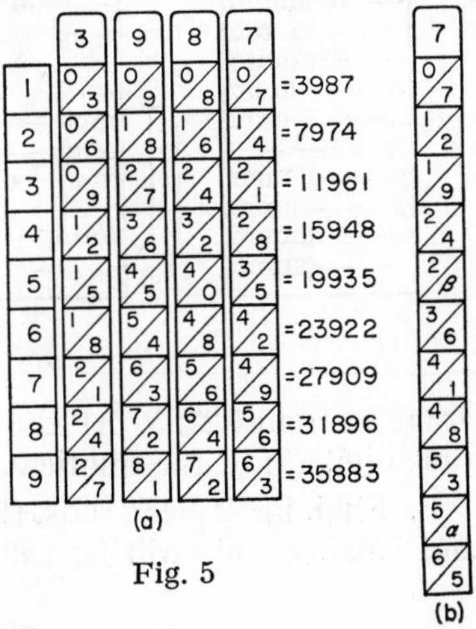

Fig. 5

For example (see Fig. 5a).

$$3987 \times 8 = \begin{array}{|c|c|c|c|} 2\!\!\diagup\!\!4 & 7\!\!\diagup\!\!2 & 6\!\!\diagup\!\!4 & 5\!\!\diagup\!\!6 \end{array} = 31\,896$$

(The tens above the sloping line are added to the units in the neighbouring box on the left.)

When multiplying together multidigit numbers, the products of the multiplicand and separate digits of the multiplier can, for convenience, be counted off immediately on the abacus (in the corresponding order).

Figure 6 shows the different stages of multiplying 3987 by 672,

I. 23922 hundreds (3987 × 6) are counted off on the abacus.

II. 27909 tens (3987 × 7) are added on to the number counted off.

III. 7974 units are added on to the number obtained.

On the other hand, in dividing say, 2225575 by 3987, it is required to:

1. Count off the dividend on the abacus (Fig. 7a)

2. Find, in Napier's rods a number, closest to 22255, but smaller (this will be 19935 = 3987 × 5) and sub-

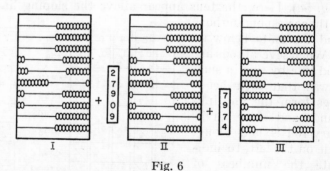

Fig. 6

tract it from 22255. At the same time, count off the first digit (5) of the quotient on the top wire (Fig. 7b).

3. Find, in Napier's rods, the number closest to 23207, but smaller (this will be 19935 = 3987 × 5).

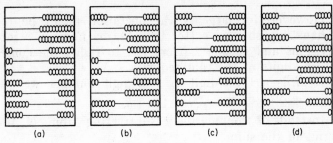

(a)   (b)   (c)   (d)

Fig. 7.

and subtract it from 23207 and count off the second digit (5) of the quotient on the second wire from the top (Fig. 7c).

4. Find, in Napier's rods, the number closest to 32725, but smaller (this will be 31896 = 3987 × 8) and subtract it from 32725, and count off the third digit (8) of the quotient on the third wire from the top.

As a net result we have the quotient required (558) on the top wires, and the remainder (829) on the bottom ones (Fig. 7d).

Make some Napier's rods out of thick cardboard or smooth plywood, and practise multiplying and dividing multi-figure numbers, using the abacus.

Two or three sets of rods should be prepared, since the multiplicand might have digits that are alike; don't forget rods with the heading "0"!

Napier's rods may be useful even in calculations in other systems of notation. In Fig 5b a rod is shown for multiplying seven by the numbers, 1, 2, 3, 4, 5, 6, 7, 8, 9, 10, 11 — in the duodecimal system of notation ($\alpha = 10$, $\beta = 11$).

The Extraction of the Square Root

It is well known, that the sum of n consecutive odd numbers equals n^2: $1 + 3 + 5 \ldots + (2n - 3) + (2n - 1) = n^2$. Therefore, the calculation of $\sqrt{N}$ (N being a natural number) can be reduced to solving the problem; what is the greatest number of terms of the form 1, 3, 5, 7, 9, 11 ..., whose sum does not exceed N. This problem is easily solved on the abacus.

However, for large N the operation might be too exhausting, so it is better to split the number N into divisions and calculate (from left to right) the consecutive digits of the root required.

If N is a three-digit or four-digit number we put

$$\sqrt{N} = 10a+b.$$

Instead of consecutively subtracting from N $10a$ terms of the sum

$$S = 1+3+5+ \ldots +(20a-3)+(20a-1) = (10a)^2,$$

and then b terms of the sum

$$S' = (20a+1)+(20a+3)+ \ldots +(20+2b-1),$$

it is simpler to subtract, from the "senior" division of N, a terms only; $1 + 3 + 5 + \ldots + (2a - 1) = a^2$ (which is equivalent to subtracting $a^2 \times 100 = S$ from N) and then subtract the terms of the sum s' (where the first term is greater than the product of a and 20 by one) from the remaining quantity.

Figure 8 shows the course of calculating $\sqrt{1369}$.

We subtract the numbers 1, 3, 5, from the first division, and we slide one bead over along the top wire for each completed subtraction.

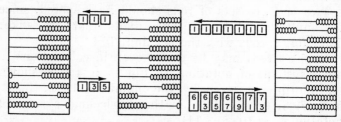

Fig. 8.

Then, on subtracting numbers 61, 63, 65, 67, ... (61 = 3 × 20 + 1), consecutively from the remainder (469), and moving one bead for each completed subtraction along the second wire from the top, we obtain $\sqrt{1369} = 37$. In order to speed up the process separate digits of the root required can be determined mentally and the sum of several terms of the form indicated can be subtracted at once from the corresponding remainder.

For example, having noted mentally that in the extraction of the root of 1369 the second digit is a seven, it is then possible, instead of subtracting the numbers 61, 63, 65, 67, 69, 71, 73 from the remainder (469), to subtract all at once $7 \times 60 + (1 + 3 + 5 + 7 + 9 + 11 + 13) = 7 \times 60 + 7^2 = 7 \times 67$, and to count off seven beads on the second top wire.

It is true, that in this way the advantage of this method is lost, and the whole process is an exact copy of the usual method of extracting the square root on paper.

If N is a 5-digit or a 6-digit number, then a in (1) is a 2-digit number. In order to find it, it is necessary to extract the root of two senior divisions of the number N and then find b in a similar manner to the above.

For instance, when calculating $\sqrt{86436}$, on subtracting the numbers 1, 3, from 8 we obtain 46436; we then subtract the numbers 41, 43, 45, 47, 49, 51, 53, 55, 57 ($41 = 2 \times 20 + 1$) from 464, and, finally we subtract numbers 581, 583, 585, 587 ($581 = 2 \times 290 + 1$) from the remainder obtained (2336). We obtain on the top wires: $\sqrt{86436} = 294$.

We shall describe one more procedure for extracting the square root of numbers, one that enables us to find directly a rational number, little different from the root required. Let

$$\sqrt{A} = a + \alpha \simeq a, \tag{2}$$

where a is the approximate value of the root required and α is the error of approximation, and it is known that $|\alpha| < \gamma$ and γ can equal $\frac{1}{10}$, $\frac{5}{10}$, $\frac{1}{100}$, etc.

From (2) we have $(\sqrt{A} - a)^n - \alpha^n$, whence, on raising the binomial to the power n, we can easily obtain, for instance:
for $n = 2$

$$\sqrt{A} = \frac{a^2 + A}{2a} - \frac{\alpha^2}{2a} = a_1 + \alpha_4, \tag{3}$$

51

for $n = 3$

$$\sqrt{A} = \frac{a^3 + 3aA}{3a^2 + A} + \frac{a^3}{3a^2 + A} = a_1 + \alpha_1, \quad (4)$$

for $n = 5$

$$\sqrt{A} = \frac{a^5 + 10a^3 A + 5aA^2}{5a^4 + 10a^2 A + A^2} + \frac{\alpha^5}{5a^4 + 10a^2 A + A^2} = a_1'' + \alpha_1''. (5)$$

If we take the first terms of these equations as the approximate value of $\sqrt{A}$, then the absolute value of the error will not exceed the numbers

$$\frac{\gamma^2}{2a}, \quad \frac{\gamma^3}{3a^2 + A}, \quad \frac{\gamma^5}{5a^4 + 10a^2 A + A^2}.$$

respectively.

For example, $\sqrt{10} = 3 + \alpha = 3; \; 0 < \alpha < 0\cdot2$ (since $3\cdot2^2 = 10\cdot24$).

From the formula (3)

$$\sqrt{10} \simeq \frac{3^2 + 10}{2 \times 3} = \frac{19}{6} = 3\cdot1666\ldots$$

(with excess, since, according to the formula (3), $\alpha_1 < 0$, always, and $|\alpha_1| < \frac{0.2^2}{2 + 3} < 0\cdot0067$), and from the formula (5)

$$\sqrt{10} \approx \frac{3^5 + 10 \times 3^3 \times 10 + 5 \times 3 \times 10^2}{5 \times 3^4 + 10 \times 3^2 \times 10 + 10^2} = \frac{4443}{1405} = 3\cdot16227758\ldots$$

(with defect, since, in this case $\alpha''_1 > 0$, and

$$|\alpha_1''| < \frac{0\cdot2^5}{1405} < 0\cdot000\,00023:$$

therefore $3 \times 16227758 < \sqrt{10} < 3 \times 16227782$.

The roots of the third, fifth, etc., powers can be extracted in a similar manner (see [9], pp. 375–378).

Addition and Subtraction In Place of Multiplication

Prior to the invention of logarithmic tables, there existed so-called *prostapheretic* tables, which were used to facilitate the multiplication of large numbers. (The name was derived from the Greek words "prostesis", meaning adding, and "aphareisis", meaning taking away.) These tables consisted of tables of values of the function $\left[\frac{Z^2}{4}\right]$ (see § 2) for natural values of Z (see [10] pp. 55–56). Since for integral a and b

$$ab \equiv \frac{(a+b)^2}{4} - \frac{(a-b)^2}{4} = \left[\frac{(a+b)^2}{4}\right] - \left[\frac{(a-b)^2}{4}\right]$$

(the numbers $a + b$ and $a - b$ are either both even or both odd; in the latter case the fractional parts of $\frac{(a+b)^2}{4}$ and $\frac{(a-b^2)}{4}$ are identical), the multiplication of a and b is reduced to finding $a + b$, $a - b$, and then finding the difference of the numbers $\frac{(a+b)^2}{4}$ and $\frac{(a-b)^2}{4}$ whose values are taken from the tables.

For the multiplication of three numbers the following identity may be used

$$abc = \frac{1}{24}\{(a+b+c)^3 - (a+b-c)^3 - (a-c-b)^3 \\ -(b+c-a)^3\}. \quad (6)$$

It follows from it that the existence of a table of values of the function $\frac{Z^3}{24}$ enables us to reduce the calculation of the product abc to the determination of numbers $a + b + c$, $a + b - c$, $a + c - b$, $b + c - a$, and from them, with the help of tables, to the finding of the right-hand-side of the eqn. (6).

An example of this kind of table is shown here for $1 \leqslant Z < 30$. In it ordinary figures represent the values of $\left[\frac{Z^3}{24}\right]$ and small figures give the values k, where

$$\frac{z^3}{24} = \left[\frac{z^3}{24}\right] + \frac{k}{24}, \quad 0 \leq k \leq 23,$$

tens \ units	0	1	2	3	4	5	6	7	8	9
0		0_1	0_8	1_3	2_{16}	5_5	9_0	14_7	21_8	30_9
1	41_{16}	55_{11}	72_0	91_{13}	114_8	140_{15}	170_{16}	204_{17}	243_0	285_{19}
2	333_8	385_{21}	443_{16}	506_{23}	276_0	651_1	732_8	820_3	914_{16}	1016_5

It is simple to obtain, using formula (6) and the table;

$$9\times9\times9 = 820_3-30_9-30_9-30_9 = 729,$$
$$17\times3\times4 = 1016_5-385_{21}-91_{13}+5_5 = 544 \quad \text{(check!)}.$$

On Calculating Logarithms

We shall also describe a simple procedure enabling us to calculate logarithms of numbers with the aid of tables of cubes of natural numbers (see [11]). Take Barlow's tables [3], which give, among others, the cubes of all 4 digit numbers, and check all calculations given below.

Let

$$\log 13 = c_0 \cdot c_1 c_2 c_3 \ldots _{(3)} = c_0 + \frac{c_1}{3} + \frac{c_2}{3^2} + \frac{c_3}{3^3} + \ldots,$$

i. e. the logarithm required is represented in the form of systematic fraction, base 3.
Then

$$10^{c_0 + \frac{c_1}{3} + \frac{c_2}{9} + \ldots} = 13,$$

therefore, $c_0 = 1$.
Dividing both sides of the equation by $10^{c_0} = 10$ and then cubing them, we obtain

$$10^{c_1 + \frac{c_2}{3} + \frac{c_3}{9} + \ldots} = 1 \cdot 3^3 = 2 \cdot 197,$$

whence $c_1 = 0$.

Rapid Calculations

Since in the exact cubing of 2·197 and in further calculations, the numbers obtained would be very unwieldy, we shall in future calculations use four significant figures only, and we shall determine the *lower* and *upper* bounds (L. B. and U. B.) of the numbers concerning us. (Whenever we round of. a number, we shall write its value with excess in the column U. B. and its value with defect in the column L. B.).

Further calculations yield:

	L. B.	U. B.	
$10^{c_2 + \frac{c_3}{3} + \ldots}$	10·60	10·61	$c_2 = 1$
$10^{c_3 + \frac{c_4}{3} + \ldots}$	1·191	1·195	$c_3 = 0$
$10^{c_4 + \frac{c_5}{3} + \ldots}$	1·689	1·707	$c_4 = 0$
$10^{c_5 + \frac{c_6}{3} + \ldots}$	4·818	4·974	$c_5 = 0$
$10^{c_6 + \frac{c_7}{3} + \ldots}$	111·8	123·1	$c_6 = 2$
$10^{c_7 + \frac{c_8}{3} + \ldots}$	1·397	1·866	$c_7 = 0$
$10^{c_8 + \frac{c_9}{3} + \ldots}$	2·726	6·498	$c_8 = 0$
$10^{c_9 + \frac{c_{10}}{3} + \ldots}$	20·25	274·4	$c_9 = 1,\ c_9 = 2$
$10^{c_{10} + \frac{c_{11}}{3} + \ldots}$	8·303	20·67	$c_{10} = 0,\ c_{10} = 1$
$10^{c_{11} + \frac{c_{12}}{3} + \ldots}$	572·4	8·832	$c_{11} = 2,\ c_{11} = 0$

Hence

$$1+\frac{1}{3^2}+\frac{2}{3^6}+\frac{1}{3^9}+\frac{2}{3^{11}} < \log 13 <$$

$$< 1+\frac{1}{3^2}+\frac{2}{3^6}+\frac{2}{3^9}+\frac{1}{3^{10}} + \cdots$$

$$1{,}113911 < \log 1{,}3 < 1{,}113979.$$

Try to calculate in this way the logarithm of this or that number and compare the result obtained with the value of that logarithm, taken from the logarithmic tables.

§ 9. Numerical Giants

In physics, chemistry, and astronomy, numerical "giants" and "dwarfs" are often encountered; for example, distances from the earth to the nearest stars are of the order of 10^{13} to 10^{14} km, the radius of an atom is of the order of 10^{-8} cm, the number of molecules in one gram-molecule of a substance equals 6×10^{23} approximately (Avogadro's number).

In order to obtain a clearer image of the dimensions of these numbers, it is useful to make various artificial comparisons. For example, the following gives a vivid picture of Avogadro's number;

If a tumblerful of water containing nothing but marked molecules were to be poured into, and mixed evenly with, the waters of the five world oceans, then every tumblerful of oceanic water would contain no less than five hundred marked molecules[26].

It is much more complicated to give a distinct representation of large and small numbers encountered in solving various mathematical problems. Let us discuss several examples:

1. log x increases with increasing x so slowly, that the inequality log $x > 100$ holds only when $x > 10^{100}$.

In order to imagine the size of this number, verify by calculation[27] that it exceeds the number of molecules of water required to fill a cube, whose edge is 70 million light years (regarding the density of water as continuing to be l, and that one cubic centimeter of water contains $\frac{1}{3}$ 1 $\times 10^{23}$ molecules).

How, then, can one picture the size of the number $K = (9^9)^9 \doteq 4 \cdot 28 \times 10^{369693099}$ (check, given that log $9 = 0 \cdot 95424250943932\ldots$) in comparison with which even the numbers of the herd in Archimedes' problem ($N \doteq 77 \times 10^{206543}$; see § 4) is a number of "ultra-ultra-microscopic dimensions".

But a number much exceeding $(9^9)^9$ is also connected with Archimedes. In his work *"Psammites"* (*The Sand-Reckoner*) Archimedes brings the classification of numbers up to the number $10^{80\ 000\ 000\ 000\ 000\ 000}$ in our notation. In Archimedes' classification — see [15] p. 13 — this number contains a myriad myriad units of the myriad-myriad order, of the myriad-myriad period; myriad = 10000.

However, even this giant of Archimedes has to yield pride of place to the apparently insignificant looking number $Q = 4^{4^{4^4}}$. Indeed $Q = 4^{4^{256}} > 4^{1 \cdot 33 \times 10^{154}} > 10^{8 \times 10^{153}}$ (2[a]). In order to give an idea about the "length" of the number Q, when written down in the decimal system of notation (but certainly not about the number Q itself) a certain book of the nineteenth century says: "Imagine a segment of such length, that a light-ray would require quintillion (10^{30}) years to traverse it. Then imagine a sphere with a diameter, equal to this segment, filled with printers' ink. All this ink would be insufficient to print this number even if the smallest type in existence were used." (see [15], p. 31–32).

Indeed, it can be established easily([28]), that the volume of this sphere is less than $\frac{1}{2} \times 10^{144}$ cm³ (check!).

Supposing that each cubic centimeter of ink is sufficient to print 1 000 000 000 digits, we could manage to print a number smaller than one containing $\frac{1}{2} \times 10^{153}$ digits but Q contains more than 8×10^{153} digits.

2. It is known that a^x (for $a > 1$), increases, with the increase of x, faster than x^n ($n > 0$) and x^n increases faster than $\log_b x$ ($b > 1$).

This means, that

$$\lim_{x \to \infty} \frac{x^n}{a^x} = 0 \quad \text{and} \quad \lim \frac{\log_b x}{x^n} = 0.$$

However, for instance, the equation $\lim f(x) = 0$, where $f(x) = \frac{x^{1000000}}{1000001^x}$ may seem paradoxical, if judged by the beginning of the table

x	0	1	2	10^6	10^7	10^{13}	10^{14}
$x^{1\,000\,000}$	0	1	$2^{1\,000\,000}$	$10^{6\,000\,000}$	$10^{7\,000\,000}$	$10^{13\,000\,000}$	$10^{14\,000\,000}$
$1 \cdot 000001^x$	1	$1 \cdot 000001$	$1 \cdot 000002$	$e \approx 2{,}718$	$e^{10} \approx 10^{4 \cdot 34}$	$e^{10^7} \approx \atop \approx 10^{4 \cdot 34} \cdot 10^6$	$e^{10^8} \approx \atop \approx 10^{43,429} \cdot 10^6$

But the ending of the table shows that for some value x in the interval $(10^{13}, 10^{14})$, $f(x) = 1$ and, since $\log f(10^{14}) \doteq -29 \cdot 43 \times 10^6$ (check) therefore $f(10^{14}) < 10^{-29 \times 10^6}$. The number e, which is to be found in the table is the base of the so-called *natural* logarithms: $\log_e N = \ln N$ and the natural and the decimal logarithms of the same number are connected by the relationship; $\ln N = \log N \times 2 \cdot 3025851$.

In higher mathematics it can be shown that

$$e = \lim_{n \to \infty} \left(1 + \frac{1}{n}\right)^n = 2 \cdot 7182818 \ldots,$$

hence $\log e \doteq 0 \cdot 4342945$ and $e \doteq 10^{0 \cdot 4342945}$.

Attempt to prove[29], that for the function

$$\varphi(x) = \frac{\log_{1 \cdot 000001} x}{x^{1 \cdot 000001}},$$

$\varphi(e^{31\,000\,000}) > 1$, but $\varphi(e^{32\,000\,000}) < 1$.

3. The factorial function $n!$ increases extraordinarily swiftly, when n increases. In order to appreciate the speed of growth of this function the inequalities (see [26]) $\sqrt{(2\pi n)}\, e^{-n} n^n < n! < \sqrt{(2\pi n)}\, n^n e^{-n} e^{\frac{1}{12n}}$ (Stirling's formula) can be used, whence

$$\frac{1}{2} \ln{(2\pi n)} + n \ln{n} - n < \ln{(n!)} <$$

$$< \frac{1}{2} \ln{(2\pi n)} + n \ln{n} - n + \frac{1}{12n}.$$

These inequalities — for large values of n — give close bounds for $\ln(n!)$. With their aid, it is easy to verify[30] that, after appropriate multiplications, the number 10 000! contains 35 660 digits, the number 100 000! contains 456 574 digits, and the number 100 000! contains 5 565 709 digits.

§ 10. Games with Piles of Objects

Three games, whose theory has been fully and exhaustively worked out, are described below. In each of them, the majority of first moves are favourable to the player making them, i. e. if he plays correctly, he is assured of winning, and only exceptional situations favour his opponent. (It is assumed, that both players are familiar with the theory of the game.) Thus the games can only serve as such for persons unacquainted with their theory.

Bachet's Game

From a pile, initially containing n objects, two players take alternately an arbitrary number of objects (but no less than one and no more than a) at a time. The winner is that player, who takes, when his turn comes, all the remaining objects. The situation is unfavourable for the player making the next move, if the number of objects in the pile (denote it by m) is divisible by $a + 1$. Indeed, when $m = a + 1$ any move by the player results in his opponent collecting all remaining objects. If $m = (a + 1)s$ (s is any natural number), then the opponent can, by making the appropriate move after any move by the player, leave $(a + 1) \times (s - 1)$ objects in the pile, then $(a + 1)(s - 2)$ objects, and so on, finally bringing the number of objects in the pile to $a + 1$, which secures his victory.

In all other initial situations (when $m = (a + 1)s + r$; $1 \le r \le a$) the first player, by taking r objects, condemns his opponent to defeat.

This game was described by Bachet in a slightly different form, as early as in 1612. Two players call out numbers from 1 to 10 and the winner is the player, who first makes up the sum of numbers called by the two players to one hundred.

Tsyanshidzi (a game with two piles of objects)

The theory of the Chinese national game *tsyanshidzi* ("picking stones") is much more complicated. Its rules are as follows.

Two players are allowed to select, from two piles of various objects, either; (1) an arbitrary number of objects from one pile (even all the objects, but not fewer than one) or (2) simultaneously the same number (also arbitrary) of objects from each pile, but no fewer than one object from each pile.

The player, who takes all remaining objects, when it is his turn to make a move — we call thus each of the above-mentioned operations — is the winner.

We shall call the situation, in which the piles contain l and k objects respectively, position (k, l) or (l, k) (the order does not matter in this case). Let us construct so-called "special" positions

$$(c_0, d_0), (c_1, d_1), (c_2, d_2), \ldots, (c_n, d_n), \ldots, \qquad (1)$$

starting from the following conditions:

(1) $c_0 = d_0 = 0$.

(2) the component c_n of the position (c_n, d_n) ($n = 1, 2, 3, \ldots$) is taken to equal the smallest of the natural numbers unused in the construction of the positions $(c_0, d_0), (c_1, d_1), \ldots, (c_{n-1}, d_{n-1})$.

(3) $d_n = c_n + n$.

The remaining positions we call *non-special*.

Here are the first two special positions (0, 0), (1, 2), (3, 5,) (4, 7), (6, 10), (8, 13), (9, 15).

Special positions possess three properties:

I. Every natural number enters one and only one position.

Indeed, taking the smallest of the numbers, not used in the preceding special positions, for c_n, we guarantee, that every natural number is bound to be in one of the positions (1). It is clear, that c_n does not coincide with any of the components of the preceding special positions, and since, for $k < n$, $d_n = c_n + n > c_k + k = d_k > > c_k$, d_n cannot coincide with any of the components of the preceding positions either.

II. Any move transforms any special position into a non-special one.

Indeed, if the move alters one of the components of the special position (c_n, d_n) only, a non-special position arises, as the unchanged component cannot be a part of two different special positions; if, on the other hand, the move carried out diminishes both components c_n and d_n to the same extent, their difference remains equal to n, while in all the other special positions (c_k, d_k) the difference between the components $d_k - c_k$ equals $k \neq n$ (see condition 3).

III. It is possible to pass from a non-special position to a special one, by means of an appropriate move.

P r o o f: given the non-special position (a, a), where $a \neq 0$, it is possible to arrive at the special position $(0, 0)$ by taking away a objects from each of the piles. If, on the other hand, the non-special position (a, b), where $a < b$, is given, the following variants are possible:

1. $a = c_k$; $b > c_k + k = d_k$.

Obviously, it is sufficient to remove $b - d_k$ objects from the second pile; this gives the special position (c_k, d_k)

2. $a = c_k$; $b < c_k + k$, i. e. $b - c_k = b - a = h < k$.

It is sufficient to remove $c_k - c_h$ from each pile; this gives the special position $(c_h, b - c_k + c_n) = (c_h, h + + c_h) = (c_h, d_h)$.

3. $a = d_k$. It is sufficient to take $b - d_k$ objects from the second pile; this gives the special position $(d_k, c_k) = (c_k, d_k)$.

It follows from the properties II and III, that after any move by the player from a special position, his opponent can reduce the matter to one of the special positions. If the opponent keeps repeating this manoeuvre, there will be obtained special positions with constantly diminishing components, until there arises the special position $(c_0, d_0) = (0, 0)$. This means, that the opponent removes the remaining objects, and thus wins. If the position is non-special, the player can, by an appropriate move, reduce it to a special one. By acting similarly every time, the player wins.

Thus, if both players proceed correctly, the player, who makes the first move secures a win, if the initial position is non-special, and cannot escape defeat if the position is special.

We note, without proof (see [25], pp. 43–52, or [32], p. 426) that, for a given k, c_k and d_k can be calculated from the formulae;

$$c_k = \left(k\, \frac{1 + \sqrt{5}}{2} \right), \tag{2}$$

$$d_k = \left(k\, \frac{3 + \sqrt{5}}{2} \right). \tag{3}$$

For example, for $k = 100$:

$$c_{100} = \left(100\, \frac{1 + \sqrt{5}}{2} \right) = [100 \cdot 1 \times 6180327 \ldots] = 161,$$

$$d_{100} = \left(100\, \frac{3 + \sqrt{5}}{2} \right) = [100 \cdot 2 \times 6180327 \ldots] = 261.$$

From formulae (2) and (3) we have: $c_k < k\, \dfrac{(1 + \sqrt{5})}{2} < c_k + 1$,

$$c_k\, 0 \times 6180327 \ldots = c_k\, \frac{\sqrt{5} - 1}{2} < k < (c_k + 1)\, \frac{\sqrt{5} - 1}{2} =$$
$$= (c_k + 1)\, 0 \times 6180327 \ldots, \tag{4}$$

$$d_k\, 0 \times 3819672 \ldots = d_k\, \frac{3 - \sqrt{5}}{2} < k < (d_k + 1)\, \frac{3 - \sqrt{5}}{2} =$$
$$= (d_k + 1)\, 0 \times 3819672 \ldots \tag{5}$$

In order to find the correct move in any given position (a, b) it is convenient to have at hand a fairly extensive table of special positions. If a table of this kind is not available, or if neither of the numbers a and b occurs in a table of special positions that is available, then the following steps should be taken: it should be discovered, which of the intervals $\left(a \dfrac{(\sqrt{5}-1)}{2}, (a+1) \dfrac{(\sqrt{5}-1)}{2} \right)$, $\left(a \dfrac{(3-\sqrt{5})}{2}, (a+1) \dfrac{(3-\sqrt{5})}{2} \right)$ contains the whole number. If some whole number k is to be found within the first interval, then $a = c_k$; if it is in the second interval, then $a = d_k$.

In both cases, it is possible to achieve a special position by applying property III. It can be easily proved[31] that one of these intervals must contain some whole number, and both these intervals cannot contain a whole number at the same time.

Nim (a game with three piles of objects)

The origin of this game is unknown; its substance is as follows: there are three piles of objects; two players each take an arbitrary number of objects (not less than one) from one of the piles, (the choice is left to the player on each occasion). The winner is the player who takes all remaining objects, when it is his turn to move. To make the theory clear it is useful to recall, that any number can be represented uniquely in the form of a sum of various powers of 2.

3.
$$17 = 2^4 + 2^0,$$
$$29 = 2^4 + 2^3 + 2^2 + 2^0,$$
$$97 = 2^6 + 2^5 + 2^0.$$

Let us call the position (h, k, l) in which the piles contain h, k, l objects respectively, a *special* one, if each of the numbers 2^s $(s = 0, 1, 2, \ldots)$ either does not occur at all in the expansion of the numbers h, k, l into sums of various powers of 2, or occurs in these expansions twice altogether.

If, on the other hand, at least one of the numbers 2^s ($s = 0, 1, 2 \ldots$) appears in the expansions of the numbers h, k, l, once, or 3 times, we shall call the position (h, k, l) non-special.

For example, the position (18, 21, 7) is special, since $18 = 2^4 + 2^1$: $21 = 2^4 + 2^2 + 2^0$; $7 = 2^2 + 2^1 + 2^0$. Obviously, for any n, the position $(0, n, n)$ (one pile removed) is a special one.

The theory of this game is based on the following theorems;

THEOREM I. *The player, whose next move takes place, while the positions are the special ones, (1, 2, 3) and* $(0, n, n)$, *is doomed to defeat.*

Indeed, for any move by the player from the position $(0, n, n)$ the opponent needs only take the same number of objects from another pile, to reduce the situation to position $(0, m, m)$, where $m < n$; these tactics will finally get him to the position (0, 0, 0) i. e. to victory.

By considering a few variants the reader can verify easily, that the initial position (1, 2, 3) also leads to the inevitable defeat of the player, who moves first.

THEOREM II. *Given any two numbers m and n, it is possible to pick (uniquely) a third number p in such a way that the position (m, n, p) is a special one.*

Indeed, it is sufficient (and necessary) to include in the number p those powers of 2 which appear once altogether in the expansions of numbers m and n, and leave out of the number p those powers of 2 that do not appear in the numbers m and n at all, or appear in both m and n.

For example, for $m = 19 = 2^4 + 2^1 + 2^0$ and $n = 37 = 2^5 + 2^2 + 2^0$, p should include 2^5, 2^4, 2^2 and 2^1, which gives $p = 54$, which together with numbers 19 and 37, forms a special position; all other numbers form non-special positions with 19 and 37.

THEOREM III. *Any move made from the special position* (k, l, m) *leads to a non-special position.*

This follows directly from Theorem II.

Games with Piles of Objects

THEOREM IV. *From any non-special position it is possible to achieve a special position by means of an appropriate move.*

In order to prove this we consider the following two cases.

1. The highest power of 2 appears either in just one of the numbers or in all three at once; then it is sufficient to diminish the greatest of the numbers to the size of the number forming a special position with the other two numbers. If, initially, two largest numbers are alike, then any of them can be diminished. (It is also possible to take away the whole of the smaller pile of objects.)

2. Suppose the highest power of 2, say 2^s, appears in the expansion of 2 numbers only. Then attention should be directed at 2^{s-1}, 2^{s-2} etc., until we come across the power of 2 (denote it by 2^r), which appears in the expansions of either one or all three of the given numbers.

To obtain a special position, it is sufficient to diminish one of the numbers in such a way, that it alters only at its "tail-end" (i. e. the part of the corresponding sum of powers of 2, which contains 2^r and lower powers of 2). Here, in accordance with the first case considered above, the number with the greatest tail-end should be diminished; if two numbers happen to have the same tail-end, either may be diminished (or take away the whole of the tail-end of the third number).

Examples:

1.
$$h = 14 = 2^3+2^2+2^1,$$
$$k = 21 = 2^4+2^2+2^0,$$
$$l = 39 = 2^5+2^2+2^1+2^0.$$

Here we have the first case; it is sufficient to diminish the number l in such a way, as to obtain the number $l' = 2^4 + 2^3 + 2^1 + 2^0 = 27$, i. e. 12 objects should be taken from the third pile.

2.
$$h = 81 = 2^6+2^4+2^0,$$
$$k = 121 = 2^6+2^5+2^4+2^3+2^0,$$
$$l = 55 = 2^5+2^4+2^2+2^1+2^0.$$

Here each of the numbers 2^6 and 2^5 appears in the expansion of the given numbers twice. Since 2^4 appears in all three numbers and the tail-end is greatest in the number k, it should be brought down to $2^2 + 2^1$, which transforms the number k into $k^1 = 2^6 + 2^5 + 2^2 + 2^1 = 102$, making up a special position with h and l.

Thus, taking 19 objects from the second pile, we arrive at the special position (81, 102, 55).

$$h = 29 = 2^4 + 2^3 + 2^2 + 2^0,$$
$$k = 58 = 2^5 + 2^4 + 2^3 + 2^1,$$
$$l = 45 = 2^5 + 2^3 + 2^2 + 2^0.$$

This is a second variant of the second case: 2^5 and 2^4 appear twice each in the expansion and the greatest tailend ($2^3 + 2^2 + 2^0$) belongs to two numbers, h and l. Therefore, the tail-end belonging to one of them can be diminished by six units, bringing it down to $2^2 + 2^1 + 2^0 = 7$, and this leads either to the special position $(h - 6, k, l) = (23, 58, 45)$ or to the special position $(h, k, l - 6) = (29, 58, 39)$. It is also possible simply to discard the tail-end of the number k, which leads to the special position $(h, k - 10, l) = (29, 48, 45)$.

It follows from the two last theorems, that:

1. The player who moves first starting from any special position is doomed to defeat, since his opponent can always create, with his next move, another special position, and by repeating this maneouvre will, sooner or later, reduce the situation to the position $(0, n, n)$ or $(1, 2, 3)$, which ensures his victory.

2. In any non-special initial position, the player making the first move has to create a special position and his victory is assured.

The game "Nim" can take also the following form: three chessmen are placed in arbitrary positions on a chess board. Two persons in turn move the chessmen spiral-fashion (Fig. 9), in the direction of square A, in each move displacing one of the chessmen by an arbitrary number of squares (two, or even three chess-

men may land in one square at a time). The game ends
when all chessmen reach *A* and the winner is the player
who ends the game by his move.

N o t e 1. An attempt can be made to construct a
theory of games with somewhat modified conditions;
e.g. in Bachet's game, we might allow the player to
take 3 to 15 objects at a move, when the number of
objects in the pile exceeds 50,
and 1 to 10 objects, when the
number of objects in the pile
does not exceed 50.

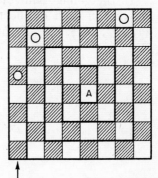

Fig. 9.

In the game with two piles
of objects, the number of
objects taken at each move
can be limited, or it can be
permitted to take objects from
both piles simultaneously only
in the ratio 1 :2, etc.

In the game "Nim" an ad-
ditional move can be intro-
duced — the taking of equal numbers of objects from
two or even all three piles.

Of course, it is not always easy to construct a theory
of a game, and often that theory is not at all simple
or elegant. However, it is highly probable that individual
cases may lead to interesting results.

N o t e 2. Even when the players are familiar with
the theory of the second or the third game, but the time
allotted to thinking out the moves is limited, the result
of the game depends on the skill of the participants.

A number of hints towards the correct conduct of
the game with two piles of objects can be found in an
article by I. V. Arnold ([38] issue 7, pp. 16–24) where
a curious connection between special positions and
Fibonacci Numbers is also pointed out.

In the game with two piles of objects try to:

a) Find correct moves for each of the following positions (27, 37), (14, 90), (47, 69), [33a].

b) Establish which of the numbers 40, 55, 140, 400 are the smaller and which the greater components of special positions, and find in each case the second component of the position[33b].

c) Construct special positions up to (c_{10}, d_{10}), making use of conditions (1)-(3) (p. 62) and also of formulae (2) and (3). Compare the results.

In the game "Nim", find the correct move (or correct moves) in each of the following positions (10, 17, 25), (47, 99, 181), (25, 43, 50), (29, 29, 18), (93, 29, 74)[34].

§ 11. Meleda

The game of Chinese origin, meleda — described as far back as the middle of the 16th century by the Italian mathematician Cardano — is depicted in Fig. 10*a*.

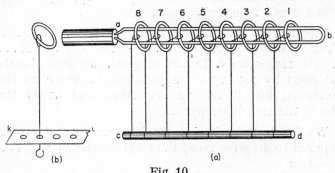

Fig. 10.

It is required to remove from a wire hairpin, *ab*, with a handle, all rings joined together by means of threads tied to a stick *cd*.

The threads can be exchanged for thin wires, and the stick *cd* for a plank *kl* with small holes, through which corresponding wires are threaded, the ends of the wires being subsequently thickened (Fig. 10*b*). The planning of the solution, which is a consequence of the constructional properties of the meleda and the recounting of all the operations needed to carry it out, adds up to an interesting mathematical problem.

On holding a meleda in one's hands (and it is quite easy to make one) it is not difficult to ascertain the following:

(1) Ring 1 can be dropped (i. e. it can be taken off the pin and passed through it downwards) or raised (i. e. it can be passed through the pin upwards and on to the pin) *regardless* of whether any other ring is on the pin or off it at the time.

(2) Any of the rings numbered 3, 4, 5, 6 ... can be dropped or raised when, and only when, a ring with the number less by one is on the pin, and all the rings with even smaller numbers are off it.

(3) Ring 2 can be dropped or raised, wherever the other rings may be, *only together with ring 1*.

From now on, the raising or dropping of any ring and also the simultaneous raising or dropping of rings 1 and 2 will be called *a move*.

Suppose the meleda has n rings; $A, B, C, D \ldots$, K, L, M, whose numbers are $n, n - 1, n - 2, n - 3, \ldots$ 3, 2, 1 respectively. In the diagram, we shall denote the rings by placing the appropriate letter above the horizontal line if the ring is raised, and below the line if the ring is dropped.

We denote by u_h $(k \le n)$ the least number of moves required to drop (or raise) the rings numbered $1, 2, \ldots$, $k - 1, k$.

The dropping of the nth ring (see I in Fig. 11) should be preceded by the situation II, which requires at least u_{n-2} moves to reach it. By dropping the ring A in the next move, we arrive at the situation III having used up $u_{n-2} + 1$ moves altogether. It is easy to confirm that starting from situation III, in order to drop ring B it is necessary to pass through the situation IV as one of the intermediate stages.

Indeed, for ring B to be dropped, ring C must be threaded onto the pin, the situation arising after the raising of C is shown in diagram V, where the ring D is in the way for dropping ring B.

Continuing to reason in this manner, we convince ourselves that the dropping of D must be preceded by

situation VI, in which the dropping of *D* is blocked by
ring *F*; in striving to drop ring *F* we must go through
the situation VII, etc.

In the final count it will be seen that the situation
IV cannot be avoided.

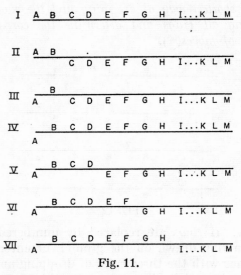

Fig. 11.

Since in the transition from III to IV at least u_{n-2}
moves are required (just as many as in the transition
from IV to III), and since to drop all the rings from the
situation IV, which can be considered as a sort of meleda with $n-1$ rings, at least u_{n-1} more moves are
required, therefore

$$u_n = u_{n-2}+1+u_{n-2}+u_{n-1} = u_{n-1}+2u_{n-2}+1. \qquad (1)$$

Obviously, $u_1 = u_2 = 1$. Making use of the recurrence
relationship (1), we obtain

$$u_3 = u_2+2u_1+1 = 4,$$
$$u_4 = u_3+2u_2+1 = 7,$$
$$u_5 = u_4+2u_3+1 = 16 \text{ etc.}$$

Applying the method of mathematical induction we
can easily find that

$$u_n = \frac{1}{2}[2^n - 1 - (-1)^n],$$

whence it follows that u_n increases rapidly with n; for example $u_{21} = 2^{20} = 1\,048\,576$.

PROBLEM. *In each of the situations shown in Fig. 12, find the shortest way of dropping and the shortest way of raising all rings and determine the corresponding numbers of moves*[35].

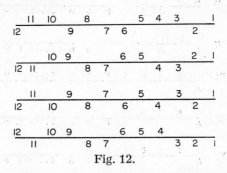

Fig. 12.

N o t e. If rings are replaced by numbered players leaving a ring drawn on the ground, or entering it in accordance with the three rules of dropping and raising rings, we obtain a livelier game, with the same principal scheme.

Such a modification of the game deserves to be noted also for the reason that, having freed oneself of the constructional peculiarities of the meleda, it is easy to alter the conditions of the game, subjecting the movements of the players to other rules, and in these altered conditions to seek the shortest solution of the problem. The construction of a wire analogue of the game with different conditions represents an additional problem, whose solution can be crowned by the creation of a new and interesting puzzle.

§ 12. Lucas' Game

In the game, invented by the French mathematician Lucas (he called it "The Tower of Hanoi") it is required to transfer n circular laminae of various sizes from column A (Fig. 13) to column B, using column C as an auxilliary one; only one lamina is to be transferred at one move (it can be moved from any column on to any other column), but it is forbidden to place a larger lamina on top of a smaller one.

It is required to indicate the shortest method of solution and to determine the corresponding number of moves u_n.

Since in order to transfer the lowest lamina to B it is necessary first to transfer (using column B as an auxilliary one) the remaining laminae on to column C, which requires no less than

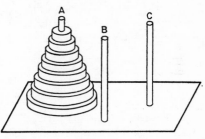

Fig. 13.

u_{n-1} moves, therefore it is obvious that $u_n = u_{n-1} + 1 + u_{n-1} = 2u_{n-1} + 1$ whence, using mathematical induction[36] it is easy to obtain $u_n = 2^n - 1$.

Readers can set themselves a number of problems of a particular or general nature, connected with Lucas' game. Let us denote the laminae in order of increasing size by numbers $1, 2, 3, 4, \ldots$. We can try to find, for example, the smallest number of moves which are

75

needed to move from the position A (8, 4, 3), B (7, 5
1), C (6, 2) — the brackets contain the numbers of the
laminae on each column, reading from below upwards —
to the position B (8, 7, 6, 5, 4, 3, 2, 1) or from the
position

$$\{A(2m, 2m-2, 2m-4, \ldots, 6, 4, 2),$$
$$B(2m-1, 2m-3, \ldots, 5, 3, 1)\}$$

or

$$\{ A(2m, 2m-1, \ldots, m+2, m+1),$$
$$B(m, m-1, \ldots, 3, 2, 1)\}$$

to the position

$$B(2m, 2m-1, 2m-2, \ldots, 4, 3, 2, 1).$$

§. 13. Solitaire

The game, which goes by the name of *solitaire* is carried out on a board with 33 squares (Fig. 14). Such a board is easily obtained by covering the four corners of a chessboard in such a way as to leave a cruciform shape (cf. Fig. 14). In Fig. 14, each square is marked by a pair of numbers, indicating the numbers of the horizontal and vertical rows, at the intersection of which the square is situated.

At the beginning of the game each square, except an arbitrary one is occupied by a draughts piece.

It is required to remove 31 pieces, given the initial empty square (a, b) and the final square (c, d) which should contain the piece surviving at the end of the game.

The rules of the game are as follows: any piece can be removed from the board, if in the next square (in the horizontal or vertical directions) there is some other piece (the remover) and if there is on the opposite side an empty square to which the remover is transfered.

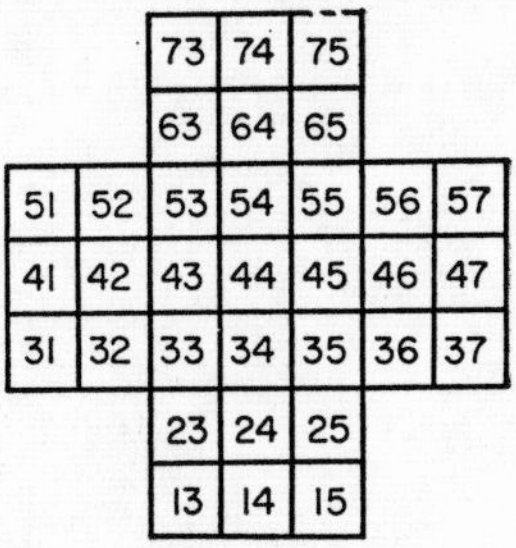

Fig. 14.

It follows from the theory of the game (see [1] or 33]) that there is a solution when, and only when, $a \equiv c$ (mod 3) and $b \equiv d$ (mod 3).

As an example let us cite the solution of the problem, when square (44) is both the initial one and the final one.

77

1. 64—44	6. 75—73	11. 65—45	16. 34—36
2. 56—54	7. 43—63	12. 15—35	17. 37—35
3. 44—64	8. 73—53	13. 45—25	18. 25—45
4. 52—54	9. 54—52	14. 37—35	19. 46—44
5. 73—53	10. 35—55	15. 57—37	20. 23—43

21. 31—33	27. 34—32
22. 43—23	28. 13—33
23. 51—31	29. 32—34
24. 52—32	30. 34—54
25. 31—33	31. 64—44
26. 14—34	

Here each move is written down by indicating the number of the starting square for the remover and the number of the square to which it moves (at the same time removing the piece from the intermediate square.

Attempt to remove 31 pieces a) when the starting square is (5, 7) and the final square is (2, 4) or b) when the starting square is (5, 5) and the final square is (5, 2) ([37]).

§. 14. The "Game of Fifteen" and Similar Games

The essence of the "game of fifteen" is as follows: a box, subdivided into sixteen squares, has fifteen numbered pieces distributed on it at random (see, e.g. I in Fig. 15). It is required to progress towards a normal arrangement III by means of a series of simple rook moves consisting each time of moving some piece to an adjacent empty square.

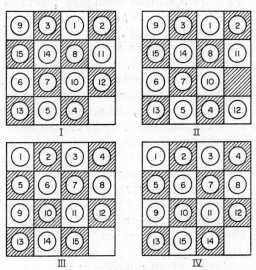

Fig. 15.

For example, by moving the piece 12 we go on from I to II; after that the pieces 10 or 11 may be moved to the empty square. Any arrangement of pieces is

79

called *a permutation*. It turns out that certain permutations are insolvable, i. e. they cannot be transformed into permutation III.

The basis of this situation rests on very simple arguments: we agree to say that two pieces are *in relative disorder* if a piece with a greater number is placed before a piece with a smaller one; for instance, the permutation III contains no inversions, but in the permutation I the piece 1 forms two inversions (with pieces 3 and 9) the piece 2 makes two more inversions with the same pieces, the piece 3 forms one inversion with 9, (the inversions of piece 3 with 2, and 3 with 1 having been accounted for already) etc.

It can be easily verified that the permutation I contains 49 inversions altogether.

We shall imagine that the empty square contains the piece 16 (fictitious), and consequently each move is reduced to a *transposition* (changing places of the fictitious piece 16 with a neighbouring piece).

In permutations I, III, IV none of the pieces form an inversion with piece 16; in the permutation II each of the pieces numbered 13, 5, 4, 12, do so.

Permutations with an even number of inversions (taking the fictitious piece 16 into account), e.g. permutations II and III, shall be called *even*, and permutations with an odd number of inversions, e.g. I and IV, shall be called *odd*.

In higher algebra there is a proof (see [20]) that the transposition of any two elements of the permutation alters its type, and consequently any move in the "fifteen" game, being a transposition of some piece with piece 16, alters the type of the permutation. (Compare for example, the transition from I to II.) It is clear that an even number of moves leads to a permutation of the original type, and an odd number of moves to a permutation of an opposite type.

If the sixteen squares are painted, for convenience, in the manner of a chessboard, then the colour of the empty box changes with each move, and therefore the following theorem holds:

THEOREM I. *All odd permutations with an empty white square and all even permutations with an empty black square are insolvable, i.e. not reducible to permutation III*

Indeed, it is only possible to progress from an odd permutation with an empty white square to a permutation with an empty square (also white) in the lower right corner in an even number of moves, i.e. only an odd permutation can be obtained here, and therefore we cannot arrive at permutation III.

The insolvability of even permutations with a black square empty can be proved similarly.

The following more general theorem is also true.

THEOREM II. *If, in addition to simple moves of the rook, we permit, in the game of fifteen, the movement of any piece to the empty square, and also the transposition of any two pieces, then no odd permutation can lead to permutation III by an even number of moves, and no even permutation can lead to permutation III by an odd number of moves.*

Indeed, when the rules of the game are altered in this manner, any move, (a transposition of two pieces, one of which may be fictitious) alters the type of permutation (prove this). But permutation III, which has to be reached eventually, is an even one.

Since a knight's move (in the same way as simple moves of the rook) causes the colour of the empty square to change alternately, theorem I holds also for the modified game of fifteen, when the pieces can move into the empty square in the manner of the knight.

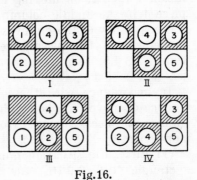

Fig.16.

THEOREM III. *Any even permutation with an empty white square and any odd permutation with an empty*

81

black square are solvable, i. e. they can be reduced to permutation III. The remaining permutations can be reduced to permutation IV.

We consider first the movement of five pieces in a rectangle of six squares. If we concentrate on the relative positions of pieces only in the clockwise direction (without taking into account the position of the empty square) then the distributions I, II, III (Fig. 16) can all be characterized by the same permutation 1 4 3 5 2 (or, which is the same, by permutations 4 3 5 2 1, 3 5 2 1 4, etc.).

It is easily seen that any horizontal move (say, the move transforming I into II) and the vertical moves in the end columns (say, the move transforming II into III) do not alter the relative disposition of the pieces, and any move in the middle column (say, the move of the piece 4 in the distribution I, leading to distribution IV) does change the relative position of the pieces, the permutation 1 3 5 4 2 characterizing the new distribution being obtained from the previous permutation by moving the appropriate number two steps: 1 3 5 4 2. By making use of this, it is easy to place any three pieces, say, 1, 2 and 3, alongside each other in the order of increasing numbers; thus, starting from the permutation 1 4 3 5 2, on moving first the piece 2, then the piece 4 in the middle column (it is necessary to create the possibility of these moves by a preliminary circular movement of the pieces) we arrive at the permutation 1 2 3 4 5. (Another starting position might have led to 1 2 3 5 4.)

In accordance with these proofs, the following sequence of moves can be employed to bring about order among the pieces, starting from any position (this does not generally give the shortest solution):

(1) First, the situation must be reached, where the rectangle "1 2 5 6 9 10" (the numbers of squares forming the rectangle are given here in accordance with the normal permutation III) contains the pieces 1 and 2 and three additional pieces, one square remaining empty.

(2) Pieces 1 and 2 are moved to their appropriate places.

(3) By a similar procedure, place the following pieces in their proper places:

3 and 4 acting within the rectangle "3   4   7   8 11 12"
5 and 6 acting within the rectangle "5   6   9 10 13 14"
7 and 8 acting within the rectangle "7   8 11 12 15 16"
9 and 13 acting within the rectangle "9 10 11 13 14 15"

(4) The pieces 10, 11, 12, 14 and 15 turn up in the rectangle "10, 11, 12, 14, 15, 16" and three of them — 10, 11 and 12 — can be moved to their proper places, which leads either to the permutation III, provided the starting position satisfied the conditions of theorem III, or to the permutation IV.

It is possible to modify somewhat the game of fifteen, by using pieces with letters spelling out some phrase printed on them. If one of the letters appears twice and each of the others once, then it is possible to arrive at a correct distribution by means of a series of moves.

Let us take the Russian phrase:* "my naveli por(ya)dok". We ascribe the numbers to the letters in the following way: 1 2 3 4 5 6 7 8 9 10 11 12 13 14 15
    m y n a v e l i p o   r (ya) d o k
and having given either 10 or 14 to each of the letters *o*, we obtain, for any initial position, two permutations

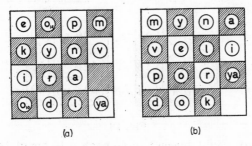

(a)   (b)

Fig. 17.

*[Translator's note: for obvious reasons this must be transliterated rather than translated. In fact, it means "we have introduced order".]

of different types, since they are transformed into each other by means of one transposition. It follows that one of the permutations must be solvable.

If we ascribe 10 to the top letter "*o*", and 14 to the bottom letter "*o*" in Fig. 17, we obtain a permutation of the even type (44 inversions), if, on the other hand, we assign the numbers to "*o*" in the reverse order, as in the diagram, we obtain an odd permutation (47 inversions).

Since the empty square in the distribution (*a*) is black, the top letter "*o*" must be transferred to the fourteenth square and the bottom one to the tenth square, in order to arrive at the distribution (*b*).

All that was said about the game of fifteen holds also for the "game of nine", where eight pieces move within a square with nine subdivisions.

Let us investigate an interesting variety of the game of nine called "chameleon". The game is played on a "board" with nine squares joined by straight lines

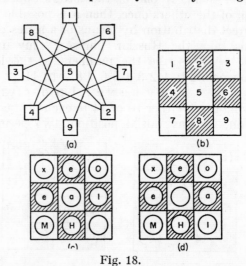

Fig. 18.

(Fig. 18*a*). Eight pieces* have each a letter of the word "(ch)ameleon" written on them and the pieces are distributed in a random fashion on eight of the squares.

* In Russian Ch is one letter

The Game of Fifteen

It is required to distribute the pieces by moving them along the lines joining the squares in such a way that, when read clockwise, starting from square 1, they should form the word "(ch)ameleon".

Having numbered the squares of the board as in Fig. 18a, it is easy to see that any two squares are joined by a straight line when and only when squares with identical numbers in diagram (b) are connected by simple moves of the rook.

Since the word "(ch)ameleon" contains two "e"s and the remaining letters appear once only (just as in the phrase used above there were two letters "o"), therefore, in accordance with the above, any initial distribution of pieces in (b) can be brought to the position (c) and therefore also to (d), which corresponds to the solution of the game (compare with (a)).

Fig. 19.

The games described in this section are one-person games. However, several persons can compete in finding shortest ways of changing from one selected position to another.

In order to complicate the game of fifteen, it is possible to forbid certain moves; for instance, it can be proved (try to do so) ([38]) that the partitions depicted in diagram 19 do not interfere with the transition from any starting position to position III or IV) (Fig. 15).

PROBLEM: *Will Theorem III, which was proved above, hold if the moves of the knight are substituted for the simple moves of the rook?* A similar question may be posed in connection with a square board containing 36 squares and 35 numbered pieces, substituting, in place of the usual knight, a "2, 3 knight" one of whose coordinates changes with each move by two units and the other by three units.

§15. Problems on Determining the Number of Ways of Reaching a Goal

Problems about Jumpers

1. *In how many ways can a man, standing in front of a row of rectangles drawn on the ground (Fig. 20), reach the n-th rectangle jumping from left to right, landing only inside the rectangles, the jumps being of any desired length.*

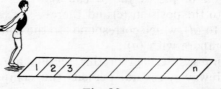

Fig. 20.

Let us denote the number of ways of reaching the sth rectangle by u_s. In solving this problem, it must be taken into account, that it is possible (but in only one way) for the jumper to jump directly to the nth rectangle, without landing in the intermediate ones. This leads to a term 1 in the expression for u_n. He can also land in k intermediate rectangles in C_{n-1}^k ways. Therefore we have the expression

$$u_n = 1 + C_{n-1}^1 + C_{n-1}^2 + \ldots + C_{n-1}^{n-1} - 2^{n-1} \qquad (1)$$

N o t e 1. Obviously, we have incidentally found the number of various representations of the number n in the form of a sum of positive integral terms (including also the case of a "sum" consisting of one term) and two representations are regarded as different, if they differ either in the terms themselves or in their order.

86

Ways of Reaching a Goal

N o t e 2. If the jumper is allowed to land an even number of times only, he has $1 + C_{n-1}^2 + C_{n-1}^4 + \cdots$ i.e. 2^{n-2} ways at his disposal.

2. *In how many ways can a man reach the n-th rectangle, if he is allowed to make only single (i.e. into the next rectangle) or double (i.e. missing out one rectangle) jumps?*

Denote the number of ways of reaching the s-th rectangle by v_s. Since the jumper can reach the rectangle numbered s from rectangles numbers $s-1$ and $s-2$ only, and to reach these rectangles he has v_{s-1} and v_{s-2} respective ways available, therefore for $s > 2$ the following equation holds:

$$v_s = v_{s-1} + v_{s-2} \tag{2}$$

It is easy to verify directly, that

$$v_1 = 1 \text{ and } v_2 = 2. \tag{3}$$

Starting from (3), it is possible to determine consecutively the values of v_3, v_4, v_5, ..., by means of the relationship (2), i.e. to represent the solution of the problem in the form of a table:

s	1	2	3	4	5	6	7	8	9	10	11	12	13	14
v_s	1	2	3	5	8	13	21	34	55	89	144	233	377	610

Equation 2 is a particular case of so-called *finite-difference* equations

$$v_{x+m} = F(v_x, v_{x+1}, \ldots, v_{x+m-1}), \tag{5}$$

studied in calculating finite differences.

If values u_0, u_1, u_2, ..., v_{m-1} are known, then it is easy, with the help of eqn. (5), also known as the *recurrence* relationship for the function v_x, to find consecutively v_m, v_{m+1}, v_{m+2}, ..., i. e. to obtain a tabular solution of equation (5). However, it would be inconvenient to try to find, say, v_{1000} in this way, and it is usual to try to represent the solution in the form $v_x = f(x)$.

It is easy to verify([39]), that eqns. (2) and (3) are satisfied by the function

$$v_n = \frac{1}{\sqrt{5}}\left[\left(\frac{1+\sqrt{5}}{2}\right)^{n+1} - \left(\frac{1-\sqrt{5}}{2}\right)^{u+1}\right]. \qquad (6)$$

By calculating the values v_3, v_4, v_5, ... with the help of this function we arrive at the same numbers as are obtained in table (4) (Check!).

An elementary discussion of the theory of linear finite — difference equations with constant coefficients:

$$v_{x+m} = a_0 v_x + a_1 v_{x+1} + \ldots + a_{m-1} v_{x+m-1},$$

is the subject of an interesting book by A. I. Markushevitch [17].

N o t e 3. The "problem about rabbits" of Leonardo Fibonacci (see [17] p. 7) also leads to the solution of eqn. (2), and as a consequence, the numbers (4) are called *Fibonacci numbers*. They possess a series of interesting properties (see [7]). We shall note here only their connection with binomial coefficients. It follows from the solution of the second problem about the jumper, that v_n equals the number of various representations of the natural number n in the form of a sum in which each term equals 1 or 2, two representations being deemed different even if they differ only in the order of the terms. On the other hand, the number of representations, in which 2 is encountered k times $\left(0 \leq k \leq \frac{n}{2}\right)$ equals C_{n-k}^k, because in this case the total number of terms equals $n - k$, and there are C_{n-k}^k ways of selecting k places out of $n - k$ places occupied by two. Therefore:

$$v_n = 1 + C_{n-1} + C_{n-2}^2 + \ldots + C_{n-\left[\frac{n}{2}\right]}^{\left[\frac{n}{2}\right]} \qquad (7)$$

N o t e 4. In the first problem about the jumper it was possible to establish the relationship

$$u_s = 1 + u_1 + u_2 + \ldots + u_{s-1}, \qquad (8)$$

whence it is easy to arrive at the equation (when $u_1 = 1$) $u_n = 2^{n-1}$.

N o t e 5. It is possible to vary the conditions in the problem about jumpers in all kinds of ways: for example, having allowed, generally speaking, single, double or treble jumps, one might add, that only single jumps are allowed from rectangles (including the starting one) with numbers divisible, say, by 5.

Denoting by w_s the number of ways of reaching the s-th rectangle in these conditions, we have, instead of one equation, a set of equations

$$\begin{cases} w_s = w_{s-1} + w_{s-2} + w_{s-3} & \text{for } s = 5k \quad \text{and for } s = 5k \pm 1, \\ w_s = w_{s-1} + w_{s-3} & \text{for } s = 5k + 2, \\ w_s = w_{s-1} + w_{s-2} & \text{for } s = 5k + 3: \end{cases}$$

here $w_1 = 1$, $w_2 = 1$, $w_3 = 2$.

Attempt([40]) to verify, by tabulating values of w_s, that in this case the jumper can reach the fifteenth rectangle in 1619 ways.

Let us now progress to problems in which the required function depends on two or more integral arguments.

The problem about the Rook

In how many ways (in the least number of moves) can a rook be moved from square (0, 0) to square (m, n), if only simple moves are employed i. e. moves to the neighbouring square either horizontally or vertically.

The numbers in brackets denote the number of the column and the number of the row respectively, at whose intersection the square is to be found, the left-hand column and the bottom row having the number 0 ascribed to them (m and n are non-negative integers).

Denote the number of ways of progressing from square (0, 0) to square (x, y) by $u_{x,y}$. Obviously, for any positive x and

$$u_{x, 0} = 1 \quad \text{and} \quad u_{0, y} = 1. \tag{9}$$

Since the rook can only reach square (x, y), when $x > 0$ and $y > 0$, either from square $(x - 1, y)$ or from square $(x, y - 1)$, which can be reached in $u_{x-1, y}$ and $u_{x, y-1}$ ways respectively, therefore

$$u_{x, y} = u_{x-1, y} + u_{x, y-1}. \tag{10}$$

We have obtained a recurrence relation for the function $u_{x,y}$, which depends on two integral (and in our problem non-negative also) arguments.

If we write in each square of the board the corresponding value of $u_{x,y}$, then, on the basis of (9) and (10), all squares of the left-hand column and bottom row can be filled by units, and then (see [10]) gradually we can fill the remaining squares by writing in numbers equal to the sum of two neighbouring numbers (the one below and the one on the left).

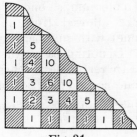

Fig. 21.

This gives us a solution (in the form of a table) of eqn. (10) given conditions (9) (see Fig. 21).

The problem about the rook can be solved in a simpler way, giving us incidentally a solution of eqn. (10) in the form of a convenient formula.

Note, that $(m + n)$ moves are required for moving the rook from square $(0, 0)$ to square (m, n) — m moves in the horizontal direction and n moves in the vertical one. Separate ways of moving the rook can be described by a scheme consisting of the letters h and v indicating the order in which horizontal and vertical moves are made.

Obviously, there are altogether $C_{m+n}^{m} = \dfrac{(m+n)!}{m!\,n!}$ various ways of selecting m places occupied by the letter h out of $m + n$ places in the scheme, which is the solution of the problem about the rook. And the solution of eqn. (10), given conditions (9) is the function

$$u_{x,y} = \frac{(x+y)!}{x!\,y!}. \tag{11}$$

A problem about a Spider

In how many ways can a spider, situated at the origin of a set of coordinates, crawl (by the shortest path) to the nodal point (k, l, m) of a space lattice?

(In a space lattice, any nodal point, i. e. a point with integral coordinates, is connected by little rods parallel to the coordinate axes to six neighbouring nodal points.) This problem is a natural generalization of the problem of the rook. If we denote the number of ways of reaching the point (x, y, z) by $u_{x,y,z}$, we have, for natural x, y, z

$$u_{x, y, z} = u_{x-1, y, z} + u_{x, y-1, z} + u_{x, y, z-1} \qquad (12)$$

To this difference equation with an unknown function depending on three integral arguments the following conditions, arising from the solution of the preceding problem, must be added:

$$u_{x, y, 0} = \frac{(x + y)!}{x!\,y!}\,; \; u_{x, 0, z} = \frac{(x + z)!}{x!\,z!}\,; \; u_{0, y, z} = \frac{(y + z)!}{y!\,z!}\,, (13)$$

Any actual method of the spider's movement from the node $(0, 0, 0)$ to the node (k, l, m) can be characterized by a sequence of letters x, y, z, indicating in what order the movements of the spider occur in the directions of the axes Ox, Oy, Oz; therefore $u_{k,l,m}$ equals the number of ways in which $k + l + m$ places can be filled by k letters "x", l letters "y" and m letters "z".

But k places out of the $k + l + m$ places available can be filled with letters "x" in $\frac{(k+l+m)!}{k!\,(l+m)!}$ ways. To each of these methods there correspond $\frac{(l + m)!}{l!\,m!}$ ways of filling l places with letters "y". Therefore, the total number of possible selections of k places in the scheme (for filling with the letters "x") and l places (for filling with the letters "y") equals

$$\frac{(k+l+m)!}{k!\,(l+m)!} \cdot \frac{(l+m)!}{l!\,m!} = \frac{(k+l+m)!}{k!\,l!\,m!}\,.$$

Thus, the solution of eqn. (12) with the limiting conditions (13) is the function $u_{x,y,z} = \frac{(x+y+z)!}{x!\,y!\,z!}$

Polydimensional Problems

The problem about a spider can be generalized to cover a four-dimensional lattice, whose nodal points are obtained from nodal points of the three-dimensional lattice by displacing them in the direction of a "fourth axis" Ou, by 1, 2, 3 etc. units of length. There is no necessity to direct the axis Ou perpendicularly to the axes Ox, Oy, Oz — for this, one would have to come out into 4-dimensional space. It is sufficient to imagine a series of 3-dimensional lattices, whose respective nodes are connected by unit rods parallel to the axis Ou.

This can be realized for three-dimensional lattices with a finite number of nodes in the form of a model, where the respective nodes of the zero, first, second, etc. three-dimensional lattices are in fact connected with each other by means of "unit wires".

In such a four-dimensional lattice each node is characterized by four integers (coordinates of the node).

Continuing the geometrical terminology the set m of numbers a_1, a_2, . . ., a_m is often regarded as coordinates of a point in m-dimensional space.

If we take the numbers a_1, a_2, . . ., a_m to be nonnegative integers, and if we regard nodes as neighbouring, if only one of their coordinates differs by a unit and all the other coordinates are identical, then the transition from node O $(0, 0, . . ., 0$ to the node A $(a_1, a_2, . . ., a_m)$, for $a_1 + a_2 . . . + a_m = n$, can be effected in n moves (progressing each time into some neighbouring node).

If we denote the moves, as a result of which only the first coordinate, or only the second coordinate, etc., increases by x_1, x_2, etc. respectively, then the number of ways of moving from node 0 $(0, 0, . . ., 0)$ to the node A $(a_1, a_2 . . ., a_m)$ in n moves equals the number of various "permutations with repetitions" of n elements:

$$\underbrace{x_1,\ x_1,\ . . .,\ x_1;}_{a_1 \text{ elements}}\ \underbrace{x_2,\ x_2,\ . . .,\ x_2;}_{a_2 \text{ elements}}\ . . .\ \underbrace{x_m,\ x_m,\ . . .,\ x_m.}_{a_m \text{ elements}}$$

By repeating the arguments given in the problem about the spider, show that this number equals $\dfrac{n!}{a_1!\,a_2!\,a_m!}$ $(^{41})$.

Filling a barrel with water can serve as an example of a multidimensional problem. Barrels numbered $1, 2, \ldots, m-1, m$ have capacities of $a_1, a_2, \ldots, a_{m-1}, a_m$ buckets respectively; in how many ways can all the barrels be filled if each full bucket empties completely into one of the barrels?

It is obvious that the number of ways equals

$$\frac{(a_1 + a_2 + \ldots\ldots\ldots + a_m)!}{a_1!\,a_2!\,a_m!}$$

A chess Problem about a King

In how many ways can a king move from the square (0, 0) to the square (k, l) if it moves in the direction of increase of one or both coordinates?*

Denote the number required by $w_{k,l}$. If the king, in its progress, makes s diagonal moves (obviously $s \leq k$, $s \leq l$) then the number of horizontal moves equals $k - s$; the number of vertical ones equals $l - s$, and the total number of moves equals $k + l - s$. Either way in which the king moves, there being s diagonal moves, can be characterized in $\dfrac{(k+l-s)!}{(k-s)!\,(l-s)!\,s!}$ ways (verify by calculating in how many ways can $k - s$ letters h, $l - s$ letters v and s letters d be arranged). Therefore, when $k \leq 1$.

$$w_{k,\,l} = \frac{(k+l)!}{k!\,l!} + \frac{(k+l-1)!}{(k-1)!\,(l-1)!\,1!} + \ldots + \frac{l!}{0!\,(l-k)!\,k!}, (14)$$

where the first term is the number of ways reaching square (k, l) without diagonal moves, the second term is the number of ways of reaching square (k, l) with one diagonal move and so on.

We suggest that the reader, starting from the identity

$$w_{x,\,y} = w_{x,\,y-1} + w_{x-1,\,y} + w_{x-1,\,y-1} \tag{15}$$

and the conditions

$$w_{x,\,0} = w_{0,\,y} = 1, \tag{16}$$

---

* When both coordinates change the move is called a diagonal move

compiles a table of values of function $w_{x,y}$, filling it with values of squares of the board, and compares the results with values obtained from formula (14).

By modifying the conditions of the problem somewhat, we may try to find the number of ways in which the king may be moved from some specific square to some other square of the chessboard in the *least number of moves*.

We split the board into zones consisting of squares, which can be reached by the king in 1 move, in 2 moves, etc. (Fig. 22), letting the solutions for, say, the first three zones be known beforehand (see numbers in the squares of these zones).

Since the square *f*5 of the fourth zone can be reached only from squares *e*4, *f*4, *g*4 of the third zone, and these squares can in turn be reached in seven, six and three ways respectively, the king can reach square *f*5 in sixteen ways (7 + 6 + 3).

Evidently, the square *g*5 can be reached in 10 (3 + + 6 + 1) ways, and the square *h*5 in four (3 + 1) ways, etc.

On filling in, consecutively, the squares of the 4th 5th, etc., zones, it is easy to verify that the king disposes of 12, 20, 266 and 357 ways of reaching squares *a*2, *a*7, *c*8 and *d*8 respectively.

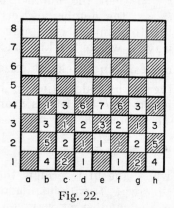

Fig. 22.

Miscellaneous Problems

There exist a number of problems in which it is difficult or impossible to construct a difference equation for the function required. This is the case, for example, if the chess knight is substituted for the king, or if in the problems about the rook and about the spider certain moves are forbidden by setting up partitions on the board, or by destroying certain rods in the space network.

In problems of this kind, the zones, all of whose squares can be reached in k moves ($k = 1, 2, 3, \ldots$) may have quite a peculiar shape, therefore the squares of different zones are either numbered or painted in different colours for convenience. Here, naturally, all squares of zone k must be revealed before proceeding to determine the $(k + l)$th zone.

In Fig. 23, the large figures mark the squares which belong to the first two zones in the problem about a knight on an unbounded chessboard; small figures indicate in how many ways the knight can reach various squares of these zones from square A.

Obviously, of the unfilled squares, those which can be reached by the knight's move from at least one of the squares of the second zone, belong to the third zone. For example, the square B belongs to the third zone, and since it is connected, via the knight's move, with 5 squares of the second zone (enclosed in a frame), the knight can reach it from square A in nine ways ($1 + 2 + 2 + 2 + 2$).

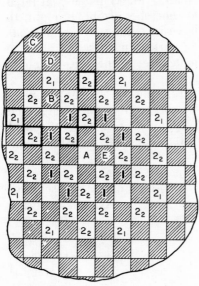

Fig. 23.

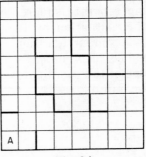

Fig. 24.

Similarly, we find that the squares, *C, D* and *E* of the third zone can be reached by the knight in 1, 6 and 12 ways.

Verify that in the presence of partitions indicated in Fig. 24, the twelfth zone in the problem of the rook, starting from A, consists of four squares, each of which can be reached by the rook in eight ways([42]).

Similar questions can be set in problems about the king and about the spider, and the set of forbidden moves can be made up in any desired manner. It is possible to aim at achieving some complicated pattern with the "*k*-zones".

It is easy to see that the fifth, sixth, etc. zones in the problem about the knight, are of a fairly regular shape, and when $k \geq 5$ the following formula holds for the number N_k of squares of the *k*th zone: $N_k = = 120 + 28\,(n - 5)$.

Consider the following questions:

1. In how many ways can the king reach the fourth zone in 4 moves, using an unbounded board([43])?

2. In how many ways can 2 (3, 4) pawns, situated in the second row of the chessboard, be brought to the eight row?([44]) (We have in mind various methods of alternating the moves of various pawns, and also the right of each pawn to make use of the initial double move or to renounce it.)

3. Try to find the general solution of the problem about the knight, i.e. to determine (at least when $k \geq 5$) the dependence of the number of ways of reaching separate squares of an unbounded chessboard on their situation on the board.

A similar question can be set, given any particular values of *p* and *q*, for the "*p, q*-knight", in whose move one of the coordinates changes by *p* units, and the other by *q* units.

§ 16. Magic Squares

We give the name of a *magic n^2-square* to a square subdivided into n^2 smaller squares, having written into them the first n^2 natural numbers in such a way that the sums of numbers in any horizontal or vertical row, and also along any diagonal of the square is equal to one and the same number $s_n = \dfrac{n(n^2+1)}{2}$. If the sums of vertical and horizontal rows only are identical, then the square is called *semi-magic*. Figure 25 shows a magic square known as *Dürer's square* after a mathematician and artist of the sixteenth century, who depicted it in the well-known painting "Melancholy". The two middle numbers of the bottom row form the figure 1514 — the date at which the painting was completed.

It is easy to investigate fully the topic of magic squares, when $n = 3$. Indeed, $s_3 = \dfrac{3(3^2+1)}{2} = 15$, and there exist only eight ways of expressing the number 15 in the form of a sum of various numbers (from 1 to 9):

$$15 = 1+5+9 = 1+6+8 =$$
$$= 2+4+9 = 2+5+8 = 2+6+7 = 3+4+8 =$$
$$= 3+5+7 = 4+5+6.$$

Note that each of the numbers 1, 3, 7, 9 enters into two of the sums given, each of the numbers 2, 4, 6, 8 into three of them, and the number 5 only comes into four. On the other hand, of the eight rows of three squares each — three horizontal ones, three vertical ones and

97

two diagonal ones — three rows pass through each of
the corner squares, four through the central square and
two through all the remaining sqaures. Therefore the
number 5 is found to be situated in the central square,
the numbers 2, 4, 6, 8 in the corner squares, and the
numbers 1, 3, 7, 9 in the remaining squares of the
larger square.

16	3	2	13
5	10	11	8
9	6	7	12
4	15	14	1

(a)

6	7	2
1	5	9
8	3	4

(b)

2	7	6
9	5	1
4	3	8

(c)

Fig. 25.

Since the numbers 2, 4, 6 and 8 can be distributed
in the corner squares so that the diagonal sums equal
15 each in eight ways only, and their position fully
determines the position of the numbers 1, 3, 7 and 9,
it can be stated that there exist only eight nine-square
magic squares. Two of them, being mirror-images of
each other are shown in Fig. 25 *b*, *c*; the six remaining
ones are obtained from these squares by rotating them
about their centres by 90°, 180°, 270°. With the increase
of n, the number N of various squares with n^2 small
squares in them, increases rapidly, and, although a
general formula expressing the dependence of N on
n has not been found yet, it has been established that
there exist 880 various sixteen-square magic squares,
and as soon as we reach $n = 7$, the number of magic
squares reaches hundreds of millions.

There are several methods of constructing magic
squares, proposed by various authors. One of the most
elegant methods is the *method of terraces*, proposed by
Bachet: n^2 numbers are written out in order (see Fig. 26,
where $n = 5$) in n rows parallel to one of the diagonals
of the square, each row containing n numbers, and the
middlemost of the numbers being placed in the centre
of the square. We leave it to the reader to prove that

the parallel transference of all parts (terraces) situated outside the square, inside and on to the opposite side of the square, leads to magic square.

Using mirror images and rotations by 90,° 180° and 270°, it is possible to obtain from each "Bachet square" seven more magic squares.

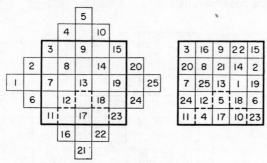

Fig. 26.

A convenient method for constructing magic squares with an even number of little squares was proposed by Ball.

Let us call operations, in which numbers α and β, α and γ, α and δ, symmetrical with respect to straight lines BB', AA' and to the centre of the square respectively, (Fig. 27a) change places, *horizontal, vertical* and *central* transpositions, and let us denote them by (α, β), (α, γ), (α, δ).

It can be seen easily that two central transpositions (α, δ) and (β, γ), where the numbers α, β, γ, δ (situated at the opposite vertices of a rectangle, whose axes of symmetry are straight lines AA', BB') change places, are equivalent to two horizontal transpositions (α, β) and (α, δ) and to two vertical transpositions (α, γ) and (β, δ) carried out consecutively.

If a square, containing $(2m^2)$ little squares, is filled with natural numbers from 1 to $4m^2$ inclusive, written out in their natural order (filling, from left to right, the top, then the second, third, etc. rows) it is easy to prove (see [25] pp. 173–176) that:

1. Both diagonals satisfy the condition for a magic square i.e. the sums of numbers in each diagonal equal $m(4m^2 + 1)$.

2. Any two vertical (horizontal) rows symmetrical about the straight line BB' (AA') satisfy the condition of magicality if any m horizontal (vertical) transpositions are carried out.

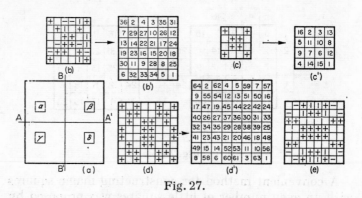

Fig. 27.

The essence of Ball's method consists of selecting transpositions as a result of which all rows and columns satisfy the condition of magicality, and the elements of each diagonal remain in the same diagonal (having perhaps changed places with each other).

Let us agree to indicate horizontal (vertical) transpositions, to which the appropriate numbers are to be subjected, by horizontal (vertical) strokes in the little squares symmetrical with respect to BB' (AA'); and to indicate two diagonal transpositions of numbers, situated in the vertices of a rectangle, by little crosses situated at the vertices of the rectangle, whose centre is at the centre of the square (these transpositions as was noted above, take the place of two horizontal and two vertical transpositions, carried out consecutively).

Figure 27, b, c, d, shows diagrammatically (for $m = 3$, 2, 4) the operations, as a result of which corresponding "naturally" filled "$4m^2$-squares" (not shown in the drawing) are transformed into squares (b'), (c'), (d').

Since, in this case, the numbers of a diagonal remain in the same diagonal, and m numbers of any horizontal or vertical row exchange places with the corresponding numbers of the symmetrically situated row, the squares (b'), (c'), (d') are magic ones.

We recommend that the reader verifies in a series of examples, that the finding of sequences of operations leading to magic squares does not present any great difficulties, and for any given m it can generally speaking be accomplished in many ways, (see, in greater detail [25], pp. 176–186).

It should be noted that for an even m the construction of a scheme for the transition from a naturally filled square to a magic one is much simpler than when m is odd, in which case the scheme must include both crosses and horizontal and vertical strokes.

When m is even, the scheme can consist of crosses alone (see, for instance, scheme d), which, however, is not a necessary condition, as can be seen, for instance, from the scheme e).

Let us widen somewhat the concept of magic "n^2-squares" by allowing that their little squares can be filled by numbers from $k + 1$ to $k + n^2$.

It is possible to look for squares satisfying various additional conditions. For example, as far back as 1544 Stifel constructed a magic "7^2-square", which, on discarding all boundary little squares yielded a magic "5^2-square" filled by natural numbers from 13 to 37, which in its turn was capable of being transformed by the same method into a "3^2-square" filled with numbers from 21 to 29 (Fig. 28a).

Figure 28b shows a *supermagic* square, i. e. one such that when an identical square is added to it on the right, all sums obtained by addition in "diagonal directions" are the same.

It is possible to construct a magic "9^2-square" disintegrating into 9 magic squares of nine little squares each (see [30] p. 106).

By analogy with magic squares it is possible to construct magic "n^3-cubes", whose subdivisions contain the first n^3 numbers (or numbers from $k + 1$ to $k + n^3$), which are distributed in such a way, that the sums of numbers in any of the $3n^2$ rows parallel to an edge of the cube, and along any of the diagonals of the cube are the same.

In [30], on pp. 108–109, the reader can find examples of magic "4^3-cube" and "5^3-cube" (it is impossible to construct a magic "3^3-cube").

It is possible to introduce the concept of magic rectangles, which give identical sums along the diagonal directions and the short rows, differing from the sums of the numbers in the long rows of the rectangle, which in turn are identical among themselves.

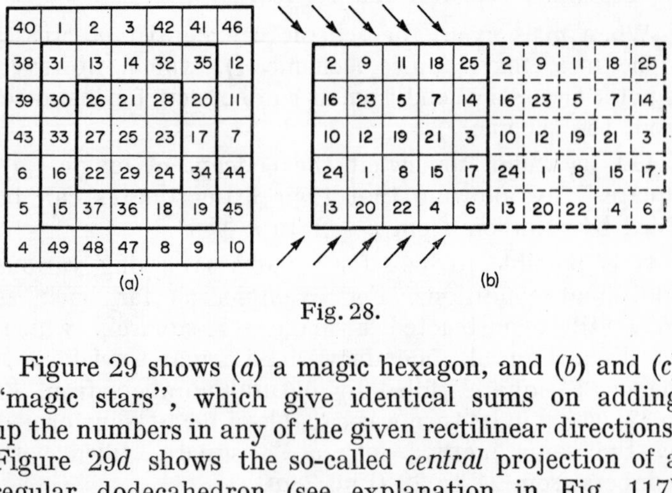

Fig. 28.

Figure 29 shows (a) a magic hexagon, and (b) and (c) "magic stars", which give identical sums on adding up the numbers in any of the given rectilinear directions. Figure 29d shows the so-called *central* projection of a regular dodecahedron (see explanation in Fig. 117) drawn inside the circle *l*. If we write down the sums of numbers situated (1) at the vertices of any face of the dodecahedron ($19 + 2 + 11 + 8 + 25$, $11 + 8 + 17 + 5 + 24$, etc), (2) along any of the dotted lines, (3) along the circumference of the circle, then each of the 19 sums equals 65.

Try to work out the following problems:

102

Magic Squares

1. Is it possible to construct a magic triangle and a magic pentagon analogous to the hexagon (*a*) by distributing in schemes (*e*) and (*f*) numbers 1 to 7 and 1 to 11 respectively?[45]

2. In the star (*b*), in addition to each quartet of numbers along its sides giving 26 when added up, the four numbers in the vertices of the great rhombi also give the same sum (12 + 1 + 7 + 6, etc.) and so do each five numbers adjacent to the vertices of the star (3 + 4 + 8 + 1 + 10, etc.).

Try to find substantially different methods of distributing numbers from 1 to 12 in the same scheme, in such a way as to get each of the fifteen groups indicated to add up to 26 again.

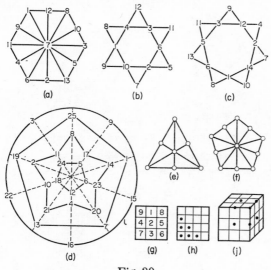

Fig. 29.

3. Distribute numbers from 1 to 9 in the little squares of a 3^2-square in such a way, that each of the four corner sums, consisting of three terms each, equals the same number *s*. For example, for square (*g*) we have $s = 14 = 4 + 7 + 3 = 3 + 6 + 5 = 5 + 8 + 1 = 1 + 9 + 4$.

In [34] it is shown, that the number n of substantially different solutions of the problem depends on s in the following manner:

s	12	13	14	15	16	17	18
n	3	6	10	9	10	6	3

When $s < 12$ and when $s > 18$ the problem has no solution.

By combining mirror images and rotations through 90°, 180° and 270°, each of the 47 substantially different solutions yield 7 more solutions, not substantially different from it.

Try to investigate the analogous problem for the 4^2-square (h) and the 3^3-cube (j) where the subdivisions containing terms of corner sums are marked in.

§ 17. Euler Squares

If the little squares of an n^2-square are filled by n elements of the first kind: $a_1, a_2, \ldots a_n$ and n elements of the second kind: $b_1, b_2, \ldots b_n$ (each of them is taken n times) in such a way, that:

(1) each little square contains one element of each kind;

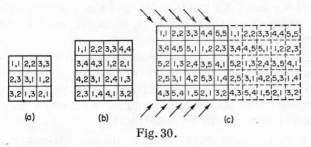

Fig. 30.

(2) each element of the first kind is combined with each element of the second kind once only;

(3) each row and each column contain all the elements of both the first and the second kind, then a so-called *Euler square* has been constructed.

If the third property applies also to the diagonals of the square, we have a *diagonal* Euler square. Figure 30 shows Euler squares for $n = 3$, 4, 5, the little squares showing the indices only of the elements of the first and second kinds.

The square (*b*) is a diagonal one, and the square (*c*) even possesses a property of this sort: if a second identical square is added on to it on the right, then in any of the ten diagonal directions, indicated by arrows in

105

the diagram, any element of the first or second kind is bound to be encountered. Such a square bears the name of an *all-diagonal* Euler square.

The problem of constructing Euler squares can be formulated differently: "Let each of the n^2 elements be characterised by its belonging to one of n classes and to one of n categories, and at the same time any two objects differ from each other either in class, or in category, or in both at the same time.

It is required to distribute the objects among the little square of the "n^2-square" in such a way as to have representatives of all classes and all categories in each horizontal and each vertical row.

For $n = 4$, we can, for instance, take the four cards, ace, king, queen and jack of the four suits, and for $n > 4$ we can take n^2 pieces of cardboard with n different diagrams on them (squares, triangles, circles, etc.,) painted in n different colours.

Euler himself tried unsuccessfully to solve a problem about 36 officers (6 officers of various ranks from each of 6 different regiments). It has been proved since that this problem is insoluble, i.e. when $n = 6$, no Euler square can be constructed.

By diminishing each number in the little squares of an Euler square by one, and then regarding the pairs of numbers as digits of numbers written in the base-n system of notation, we can from any Euler square obtain a semi-magic square, and from any diagonal Euler square a magic square.

We suggest that the reader obtains, in the same way, a super-magic 5^2-square from the all-diagonal square (e).

Any Euler square can be considered as a combination of two *Latin* squares, i.e. squares filled by n elements (each taken n times) in such a way as to encounter any one of these elements in each row and in each column. Here there should be combinations of each element of the first Latin square with each of the n elements of the second one.

§ 18. Pastimes with Dominoes

As is well known, each domino is divided into two halves, which contain *combinations with repetitions* of numbers 0, 1, 2, 3, 4, 5, 6, marked in by dots. The domino whose halves contain k and l dots shall be denoted by (k, l).

By laying the dominoes with their identical halves next to each other we can construct a "chain". A chain with identical halves at its ends can be closed. By breaking a closed chain at this or that point we can obtain open chains.

We may consider a "generalized domino set" as one on the halves of whose dominoes all possible combinations with repetitions of numbers 0, 1, 2, ..., $n - 1$, n appear.

Try to solve the following problems:

1. Show[47] that the number of dominoes and the sum of all points of a generalized domino are equal to $\dfrac{(n+1)(n+2)}{2}$ and $\dfrac{n(n+1)(n+2)}{2}$ respectively.

2. Prove[47] that when n is even, and one domino (a, b), where $a \neq b$, is removed from the complete set, the remaining dominoes can form an open chain only, ending in halves containing a and b points respectively.

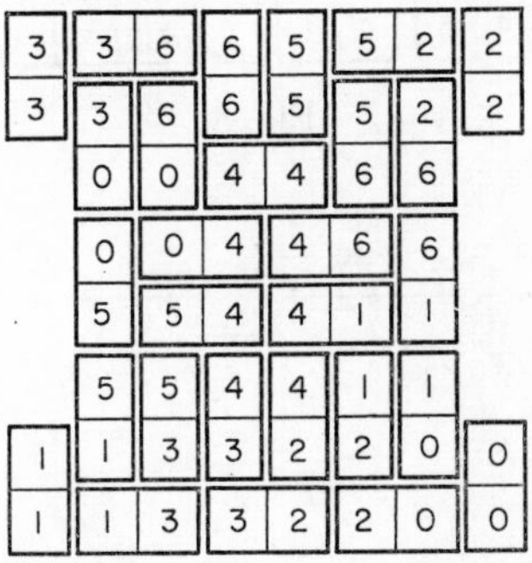

3. Prove[48] that when n is odd, a complete set of

Fig. 31.

dominoes cannot be arranged in a chain and that the longest chain contains no more than $\dfrac{n^2 + 2n + 3}{2}$ dominoes.

4. Distribute 28 dominoes among 4 players in such a way, that the second and the third player do not get a chance to put down a single domino until the first player finishes the game.

5. The diagram which appears in Fig. 31 can be subdivided into 14 squares, each containing 4 identical numbers. By interchanging numbers 0, 1, 2, 3, 4, 5, 6 and making use of mirror reflections in the vertical axis of the diagram, it is possible to obtain 10 080 ($= 2 \times 7!$) diagrams (including the given one) not differing significantly among themselves.

Could you manage to find a diagram of the same kind with a significantly different subdivision into fourteen squares each containing 4 identical numbers?

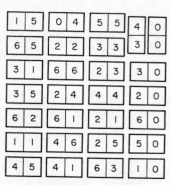

Fig. 32.

6. In Fig. 32 the dominoes are set out in such a way that on discarding mentally the right-hand column of zeros, we obtain a 7^2-square, in which the sum of points along any diagonal and in any horizontal or vertical row equals twenty-four.

Could you succeed in constructing an analogous $(n - 1)^2$ - square for a generalized domino set?

108

§ 19. Problems Connected with the Chess Board

In § 15, we considered problems in which it was required to determine the number of ways of moving a chess piece from one square of the board to another.

Let us investigate two more classical problems, those of the queens and of the knight, and a number of topics related to them. Some of them are interesting because of their close connection with the theory of sets, others may serve as a source of research and of formulating original solutions. Many interesting problems about queen or knight can be found in a book by *L. Y. Okunev* [21].

A Problem about Rooks

In how many ways can n rooks be distributed on an "n²-board" in such a way, that they do not threaten each other?

Obviously, the rooks must be placed in different rows and different columns of the board. Any such distribution of n pieces on a chess board can be characterized by a permutation of the numbers $1, 2, 3, \ldots,$ $(n-2), (n-1), n$, if we let the consecutive numbers of the permutation denote the numbers of the rows occupied by pieces in the first, second, etc., columns respectively.

Since any permutation corresponds to a definite solution of our problem, and different permutations obviously correspond to different solutions, the total number of solutions of the problem under discussion equals $n!$

The problem about rooks gains considerably in complexity if we become interested only in those solutions in which no rook is placed on the diagonal joining the lower left-hand square and the upper right-hand square. The total number of such solutions, N, is determined from the formula

$$N = n!\left(\frac{1}{2!} - \frac{1}{3!} + \frac{1}{4!} - \ldots + \frac{(-1)^n}{n}\right)$$

(see [21] p. 13).
it equals the number of permutations of n elements, in which no element occupies its normal place.

By diversifying the imposed limitations (for example, by demanding that the rooks be placed on white squares only or by making both diagonals a "forbidden zone", etc.) it is possible to arrive at combinatorial problems of greater or less degree of difficulty.

A Problem about the Queens

In how many ways can n queens be distributed on an "n^2 board" in such a way that they do not threaten each other?

Queens, situated in squares (p, q) and (s, t), threaten each other diagonally, when and only when $|p - s| = |q - t|$, here p, s are the numbers of columns and q, t are the numbers of rows at whose intersection the squares (p, q) and (s, t), respectively, are situated.

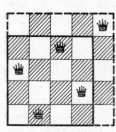

In the corresponding permutation (see problem about the rooks) the numbers q and t should be occupying the pth and the sth place respectively. A permutation of n numbers characterizes a particular solution of the problem, when the difference of the place-numbers of places occupied by any two numbers is not the same as the difference of the numbers themselves in absolute value.

Fig. 33.

It is easy to verify that, when $n = 2$, and when $n = 3$, the problem about the queens is insoluble. The solubility of the problem, when $n = 4$ and when $n = 5$ is easily verified from Fig. 33; and for $n \geq 6$ the solubility of the problem is seen from the examination of permutations given in the following table:

Form of numbers	Permutation, giving solution of problem about queens
$6k$ $6k+4$	$2, 4, 6, 8,..., n-2, n, 1, 3, 5, 7,..., n-3, n-1$
$6k+2$	$4, n-2, n-4, n-6, ..., 10, 8, 6, n, 2, n-1,$ $1, n-5, n-7, ..., 5, 3, n-3$
$6k+1$ $6k+5$	$n, 2, 4, 6, ..., n-3, n-1, 1, 3, 5,...,n-4, n-2$
$6k+3$	$n, 4, n-3, n-5, ..., 10, 8, 6, n-1, 2, n-2,$ $1, n-6, n-8, ..., 7, 5, 3, n-4$

Attempt to prove that in each of the permutations shown the difference of any two numbers differs in absolute value from the difference of their place-numbers (see [21] pp. 23–25). Verify this, say, for $k = 1$ and for $k = 2$.

The question of the number of solutions of the problem about queens on an "n^2-board" for an arbitrary n has not been resolved yet, in spite of the efforts of numerous mathematicians. In a particular case ($n = 8$), it has been established, that the problem has 92 solutions. Among those, there are only 12 "independent" solutions; they are characterized by the permutations

15863724	25741863	27581463
16837425	26174835	35841726
24683175	26831475	36258174
25713864	27368514	35281746

The remaining solutions are obtained from these twelve by means of the "reflection of all pieces" in the vertical midline of the chessboard, and by means of rotating the pieces (taken in the "initial" and the "reflected" positions) through 90°, 180°, 270° with respect to the centre of the chessboard.

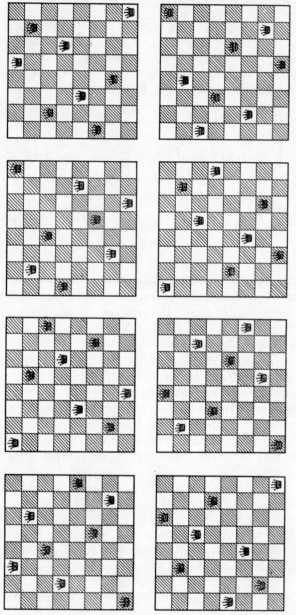

Fig. 34.

In Fig. 34 the upper row gives a solution corresponding to the permutation 15863724, and three solutions obtained from it by rotation by 90°, 180° and 270°; in the lower row, there are to be found the "reflected" solution and solutions obtained from it by rotating through the same angles as before. Thus, each of the permutations shown gives, generally speaking, eight solutions altogether; the exception is provided by the last permutation, which gives three additional solutions only.

The two problems considered can be generalized in various directions. For example, n rooks or n queens may be distributed on an "m^2-board" ($m > n$) or on a "p, q-board" ($p \geq n$, $q \geq n$).

Solutions of the problem may be sought, in which, say, no two of the pieces set out are connected by the knight's move, etc.

Finally, a rectangular, three-dimensional "n^2-network" (or even a k-dimensional "n^k-network") may be taken, extending to $n - 1$ units in the direction of each dimension, and n^2 (n^{k-1} respectively) rooks or queens, not threatening each other, distributed at its nodes.

Here we assume that the rook's move and the queen's move takes us from any given nodal point to any nodal point only one of whose coordinates differs from the corresponding coordinate of the initial node, and the queen's move can take us also into nodal points, two of whose coordinates differ from the corresponding coordinates by numbers having equal absolute values. The queen's move can, of course, be defined differently, e. g., by supposing that nodes $A(a_1, a_2, \ldots, a_k)$ and $B(b_1, b_2, \ldots, b_k)$ are connected by the queen's move, if the differences $b_s - a_s$ (when $s = 1, 2, \ldots, k - 1, k$), without depending on each other, take on one of three values: 0, m, $-m$ (m being an arbitrary integer), at least one of the differences being other than zero.

When the queen's move is treated in this way, the question of determining the smallest values of n (for various k), for which the problem about queens is soluble, acquires some interest.

113

Note that the problem about distributing n bishops (situated on both white and black squares) on an n^2-board turns out to be intermediate in difficulty between the problems about the rooks and about the queens. The number of its solutions was established comparatively recently by the Soviet mathematician S. E. Arshon ([38], 8th edition, pp. 24–29.).

A Problem about a Knight

Using the knight's moves, visit each square of the chessboard once and once only.

This problem drew the attention of many prominent mathematicians, who suggested a series of procedures towards obtaining some of its particular solutions (see [21], pp. 54–74). To this day, however, it has not been established what is the total number of solutions of this problem, although it is known that that number is very great.

The usual chessboard may be given up in favour of an n^2-board ($n \neq 8$) or a rectangular m, n-board. Finally, simple rectangular boards can be exchanged for boards with blank spaces inside them.

Figure 35 shows several solutions of the problem about the knight, featuring bizarre patterns obtained on joining up, by means of straight lines, the centres of the squares consecutively visited by the knight in its journey over the chessboard.

The Russian chess-player of the 19th century, Yanich, found a solution of the problem about the knight, leading, by numbering the squares in the order in which they are visited, to a semi-magic square, with equal sums along the rows and along the columns ($s = 260$) (Fig. 35 c).

A "p, q-knight", moving each time p squares in one direction and q-squares in the other, can be substituted for the usual knight, p and q being numbers of which one is even and the other odd, otherwise the p, q-knight would have to move in squares of one colour only.

It is easy to verify, that a 3^2-board cannot be covered by the moves of an ordinary knight. The same applies to a 4^2-board although on a rectangular 3, 4-board the problem is soluble.

It is natural, that with the increase of the numbers p and q (or at least one of them) the problem may become insoluble on an n^2-board, even for some values $n > 4$.

It is interesting to investigate, for instance, whether an n^2-board (for $n = 5, 6, 7, 8$) can be covered by a 2, 3-knight and 1, 4-knight.

In those cases in which the problem turns out to be insoluble, the question can be posed about the minimal number of squares left out in the coverage (for instance, an ordinary knight can visit all squares of a 3^2-board, except the central one). In the soluble cases, solutions with original patterns may be sought.

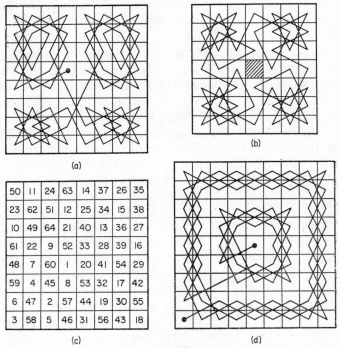

(a)

(b)

50	11	24	63	14	37	26	35
23	62	51	12	25	34	15	38
10	49	64	21	40	13	36	27
61	22	9	52	33	28	39	16
48	7	60	1	20	41	54	29
59	4	45	8	53	32	17	42
6	47	2	57	44	19	30	55
3	58	5	46	31	56	43	18

(c)

(d)

Fig. 35

115

The usual problem about the knight is equivalent to the following arithmetical problem; write down 64 pairs of integers of various kinds; (a_1, b_1), (a_2, b_2), ..., (a_{64}, b_{64}). (numbers a_k and b_k can have values from 1 to 8) such, that for any two neighbouring pairs the following condition would hold

$$(a_{k+1} - a_k)^2 + (b_{k+1} - b_k)^2 = 5$$

(one component changes by one unit, and the other one by two units). If, in addition $(a_{64} - a_1)^2 + (b_{64} - b_1)^2 = 5$, it is possible to get from the final point to the starting point by means of the knight's move. A coverage of the board, satisfying the latter equation is called *closed*.

Miscellaneous Problems

Problems, involving the coverage of the board by the king or by the rook (simple moves of the rook) are also of interest.

Since a rook, in any position, has at his disposal, four moves at most, while the king and the knight can make as many as eight different moves from certain squares, it is natural to expect that problems involving the rook are the simplest ones.

Could you find how many different ways are available to the rook for covering an n^2-board, when $n = 3$, 4, 5, ... or for covering an m, n-board for various values of m and n (perhaps you may even succeed in finding a general solution of this problem, for any m and n).

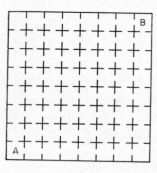

Fig. 36.

It should be noted, that for odd values of the product mn, there do not exist any closed coverages of an m, n-board by a rook, since the starting square and the final square are necessarily of the same colour (the rook must make an even number of moves in order to reach the final square) and they cannot be spanned by a simple move of the rook.

It is also obvious that, for boards with an even number of squares, the starting square and the final square cannot be (for the rook) of the same colour, which explains, in particular, the insolubility of the following problem: starting from room A (Fig. 36), visit all rooms once and finish your circuit in room B.

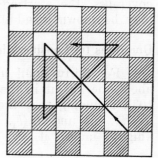

Fig. 37.

It remains to take note of the problems in which this or that chess-piece must pass through all squares of the board, but does not have to stop in each, and perhaps passes through certain of them several times. It is easy to verify, for instance, that the rook can cover an n^2-board in $2n - 1$ moves. For a 3^2-board the substitution of a queen for a rook does not diminish the number of moves required to cover the board. However, if a square area of nine squares is situated inside a large board, and if the queen is allowed to cross its boundaries, the whole area can be covered in four moves (Fig. 37) and not in five as is the case with the rook. A similar question can be set for $n > 3$, for an n^2-area situated inside a large board.

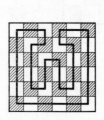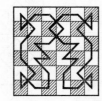

Fig. 38.

It is interesting to seek solutions of the problem about the knight (king or rook), where the chess-piece follows a broken-line circuit (see Fig. 38). Here, for nonclosed coverages, the starting square and the final square may be shown in addition.

Questions for Consideration

1. For fairly small n and k, consider the problem about the knight in the case of an n^k-network (see p. 103), assuming that the nodes of the network, $A(a_1, a_2, \ldots, a_k)$ and $B(b_1, b_2, \ldots, b_k)$ are connected by knight's moves, if

$$\sum_{i=1}^{k} (a_i - b_i)^2 = 5 \text{ and } \sum_{i=1}^{k} \left| a_i - b_i \right| = 3.$$

Consider an analogous question for the "p, q-knight", when p and q are fairly small (see p. 104).

2. A bishop can pass through all similarly coloured squares of an 8^2-board in 17 moves (Fig. 39a). What is the least number of moves, in which it can pass through all similarly coloured squares of an n^2-board ($n = 9$, 10, 11, . . .)?

3. By moving alternately white and black bishops (Fig. 39b) make them change places in 36 moves, in such a way, that at no time do the bishops of differing colours threaten each other.

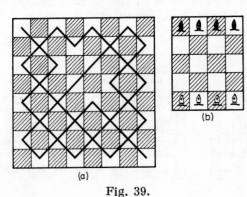

(a)

(b)

Fig. 39.

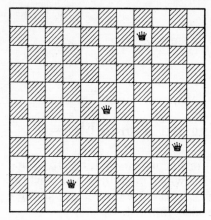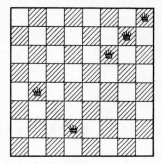

Fig. 40.

4. Show that there exist only 8 ways of covering a 3, 4-board by the knight's move and that all these ways are open([51]).

5. A chess-piece, whose move connects two squares, (k, l) and (m, n), of a chessboard, for which $(k - m)^2 + (l - n)^2 = 25$ posesses an interesting property. This piece moves both as a rook (over four squares into the fifth one) and as a 3, 4-knight; it can make exactly 4 moves from any square of an 8^2-board.

Would the reader succeed in creating other pieces, posessing an analogous property on an n^2-board ($n = 6$, 7, 9, 10, ...) or in an n^2-network ($k = 3, 4, ...$)?

6. In Fig. 40 it is seen that 5 queens in an 8^2-board can keep all squares within their striking distance, and in an 11^2-board they can keep so all squares, except those occupied by the queens themselves. Find the greatest values of N_1 and N_2 on condition that n queens ($n = 6, 7, 8, 9, ...$) can keep within striking distance all squares of an $N_1{}^2$-board or all squares of an $N_2{}^2$-board, except the ones occupied by the queens themselves.

§ 20. Making up Timetables

Let us discuss a few problems on making up time-tables, in accordance with which members of a certain collective can be grouped together, while certain supplementary conditions are also fulfilled.

1. *13 children are required to do 6 different exercises,*

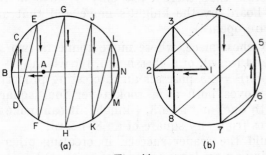

Fig. 41

while standing in a circle. Can this be done in such a way that each child has new neighbours on each occasion?

Substituting letters from A to N for children, we put A on the polygon, whose vertices divide the circle into 12 equal parts (Fig. 41a) and we distribute the remaining letters evenly along the circumference.

On moving along the polygon in the initial position, shown in the drawing, and on rotating the polygon about the centre (A moves with the polygon, all the other letters stay in their places) through 30°, 60°, 90°, 120°, and 150°, we obtain six sequences of letters:

1) $ABCDEFGHIKLMNA$, 2) $ADBFCHEKGMINLA$
3) $AFDHBKCMENGLIA$, 4) $AHFKDMBNCLEIGA$
5) $AKHMFNDLBICGEA$, 6) $AMKNHLFIDGBECA$

in which any letter has all other letters as neighbours at some time.

Try to prove that this procedure can also be used in the general case, when $2n + 1$ children have to carry out n exercises. Try to find a simpler method of solving this problem.

2. *Make up a timetable of a chess tournament with eight participants.*

We apply a method, similar to the one used in the previous problem: having placed No. 1 in the centre of the circle, and the remaining seven numbers at the vertices of an inscribed septagon, we make up a time-table for two days with the aid of a polygon, whose sides are drawn thick and thin alternately (Fig. 41*b*), by joining together first the numbers at the ends of the thick lines, and then the numbers at the ends of the thin lines;

| 1st day; | 1, 2 | 4, 7 | 6, 5 | 8, 3; |
| 2nd day; | 3, 1 | 2, 4 | 7, 6 | 5, 8. |

If we keep to the given direction of going round the circle, and if we assume that the first number of each pair plays white and the second black, then every player will get a chance of playing both white and black in the course of the two days.

By rotating the polygon about the centre, in a clockwise direction, through $\frac{4\pi}{7}$, $\frac{8\pi}{7}$ and $\frac{12\pi}{7}$ radians respectively, and using the thick lines only in the last case, it is easy to make up a timetable for the remaining five days:

3rd day	1, 4	6, 2	8, 7	3, 5
4th day	5, 1	4, 6	2, 8	7, 3
5th day	1, 6	8, 4	3, 2	5, 7
6th day	7, 1	6, 8	4, 3	2, 5
7th day	1, 8	3, 6	5, 4	7, 2

The method shown is applicable, when n (the number of players) is even; if n is odd, it is sufficient to introduce a fictitious player, and the participant who is to meet the fictitious player is regarded as free for that day. For example, by diminishing all numbers in the timetable obtained above by 1, and regarding the player No. 0 as the fictitious one, we obtain a timetable in which each of the 7 players has 3 games with the white pieces and 3 games with the black and has 1 day free of play.

3. *Fifteen children play every day in groups of three. Make up a weekly timetable in such a way, that every child has different partners each day.*

This problem, proposed by Kirkman in 1850 (in a somewhat different form) drew the immediate attention of a number of important mathematicians. Here is one of its possible solutions:

1st day:	a, b, c	d, e, f	g, h, i	j, k, l	m, n, o
2nd day:	a, d, g	b, e, h	c, l, o	j, n, i	i, k, f
3rd day:	a, j, m	b, k, n	c, f, i	d, h, o	g, e, l
4th day:	a, i, o	b, d, j	c, e, k	g, n, f	i, h, l
5th day:	a, f, l	b, g, m	c, h, n	d, i, k	j, e, o
6th day:	a, h, k	b, f, o	c, g, j	d, l, n	m, e, i
7th day:	a, e, n	b, i, l	c, d, m	g, k, o	j, h, f

Certain authors produced methods of solving analogous problems for $n = 5 \times 3^k$, for $n = 3^k$, for $n = 63 \ (= 2^6 - 1)$ and for $n = 255 \ (= 2^8 - 1)$.

It would be interesting to find, for the basic problem and its variants, easily remembered timetables, enabling the participants themselves to realize the necessary groupings in a simple way.

Sylvester, the English mathematician of the 19th century, posed the problem (apparently still unsolved) about the distribution of all possible combinations 3 at a time (their number $= C_{15}^3 = 455$) in 13 complexes, each of which on being subdivided into seven groups would give a solution of Kirkman's problem.

The solution of this problem is equivalent to the making up of a quarterly (13 weeks) timetable, in which no combination repeats itself.

Many interesting details connected with Kirkman's problem and its generalizations are to be found in [33] v. II, pp. 97–117.

In the problems discussed and in similar problems, we may try to find the number of significantly different timetables, regarding two timetables as having no significant difference if one can be obtained from the other by means of some substitution (interchange of one element by another), carried out throughout the timetable, and by means of separate days changing places.

Consider the following problem.

PROBLEM. *Make up a timetable (for different values of n) for n + 1 days so that n² schoolchildren are separated into classes of n children at a time, and any child has different classmates on each occasion.*

There exists a simple method of solving this problem, when n is a prime number, but this method is not applicable, when n is a compound number ([33] II, pp. 94–96).

§ 21. The "Problem of Josephus Flavius" and Similar Ones

Suppose that n elements are arranged in a circle, then they are counted, and every kth element is removed (if the problem is being solved on paper, the element is crossed out, if it is being solved by having objects distributed in a circle, then it is done by laying the object aside); having removed one element, the count is resumed from the next surviving element, etc.

The following questions can be set here:

1. *Which element is removed on the sth count* $(1 \leqslant \leqslant s \leqslant n)$? This was the problem, when $n = 40$ and $k = 3$, that had to be solved according to tradition, by the historian Josephus Flavius, for $s = 39$ and for $s = 40$, in order to "survive", together with his friend, after 38 removals (see [25] pp. 122–123).

2. *How should n elements be distributed, so that they are removed in a prearranged order?*

In order to solve the latter problem it is sufficient, having written out a row of the first n natural numbers standing in for the given elements, and moving from left to right, to underline every kth number, and to indicate underneath it in which turn it was removed. Here, on reaching the extreme right surviving number, we continue the count from the left end, which is equivalent to a movement round a circle.

Let us find, say, in what order should 9 cards of one suit be arranged, so that they can be laid out in order of value, from the ace to the six, by transferring one card after another from the top of the pack to its bottom, and laying every fourth one on the table.

The Problem of Josephus Flavius

Suppose the cards in the original pack are numbered from top to bottom

1	**2**	**3**	**4**	**5**	**6**	**7**	**8**	**9**
(9)	(8)	(3)	(1)	(6)	(5)	(7)	(2)	(4)

The numbers in brackets are obtained as follows: we move from left to right along the top row; we mark off every fourth number and we write its ordinal number underneath it in brackets. In further moves from left to right the marked numbers are no longer taken into account (the corresponding cards are laid on the table).

Since it is required that the first card to be laid down is an ace, it should be placed fourth from top, the king should be in the eighth place, the queen in the third, etc., and finally the six in the first place on top. Therefore, the order of the cards should be as follows; six, seven, queen, ace, nine, ten, eight, king, jack.

For large values of n, especially if we are interested in the position of one particular element, which should be the sth removed, there exists a simpler method, one that does not require the predetermination of the positions of the elements removed previously. (See [25] ch. IV).

Denote by $\{x\}$ the smallest integer satisfying the inequality $\{x\} \geq x$ and call the sequence

$$a_1 = \{a\}, \quad a_2 = \{a_1 q\}, \quad a_3 = \{a_2 q\}, \quad \ldots, \quad a_n = \{a_{n-1} q\}, \ldots$$

an *integral geometric progression* with the common ratio q.

In order to find the number t of the element to be removed sth (the initial number of elements being n, and the element removed each time being the kth) we must construct an integral geometrical progression, in which $a = k(n - s) + 1$ and $q = \dfrac{k}{k-1}$; if the greatest of the terms of this progression, which do not exceed the number nk, is denoted by A, then $t = nk + 1 - A$.

In the case of the problem about the nine cards quoted above, let us find, say, the number of the card which is the fifth one to be laid on the table. Here $n = 9$,

$k = 4$, $s = 5$, $q = \dfrac{k}{k-1} = \dfrac{4}{3}$, $nk = 36$. Therefore

$$a_1 = \{4\,(9-5)+1\} = 17, \quad a_2 = \left\{17 \times \dfrac{4}{3}\right\} = 23,$$

$$a_3 = \left\{23 \times \dfrac{4}{3}\right\} = 31, \qquad a_4 = \left\{31 \times \dfrac{4}{3}\right\} = 42 > nk:$$

therefore, $A = 31$ and $t = 36 + 1 - 31 = 6$, i.e. the fifth card to be laid on the table is the sixth card of the pack.

Solve the following problems:

1. Establish([51a]) by two methods, that Josephus Flavius and his comrade had to occupy the 13th and the 28th places respectively, in order to be removed last and one-but-last.

2. Arrange 36 playing cards in such a way, that on transferring 5 cards to the bottom of the pack and laying the sixth one on the table, there would appear in order of seniority, successively, all cards of one suit, then of the second, third and, finally, the fourth suit([51a]).

Having constructed an integral geometric progression, determine in which place should the ace of the third suit be found ($s = 19$); find the same for the jack of the fourth suit ($s = 31$) and the seven of the second suit ($s = 17$).

§ 22. Pastimes Connected with Objects Changing Places

In the first four problems of this chapter it is required to arrive at a given arrangement of objects by moving the objects according to definite rules. Some of these problems can serve as a basis for further investigations and generalization.

At the end of the chapter a problem is considered, in which a second application of a certain operation leads, in the final count, to the initial distribution of objects. The determination of the number of operations required for that purpose is connected with the properties of so-called substitutions (see p. 118) and the solution of the problem in a general form requires the aid of the theory of numbers.

The transposition of objects in pairs

In draughts, four black pieces and four white ones are arranged alternately in a straight line. According to the rules of the game we may transfer any two neighbouring pieces (without changing their relative positions and without separating them) to a new position along the same straight line. The purpose of the game is to rearrange the pieces in four moves in such a way that four black

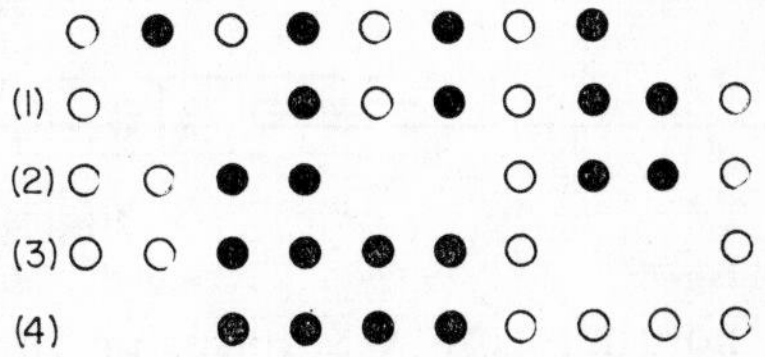

Fig. 42.

pieces appear on the left and four white pieces on the right. (The solution is shown in Fig. 42.) This problem can be generalized in the following direction: There are k kinds of pieces, s of each kind. Investigate, for what values of k and s it is possible to change over from the arrangement (a) to the arrangement (b) (Fig. 43) by means of a number of moves (in [25] pp. 189–191, it is proved that the problem is soluble for any $s > 4$ if $k - 2$).

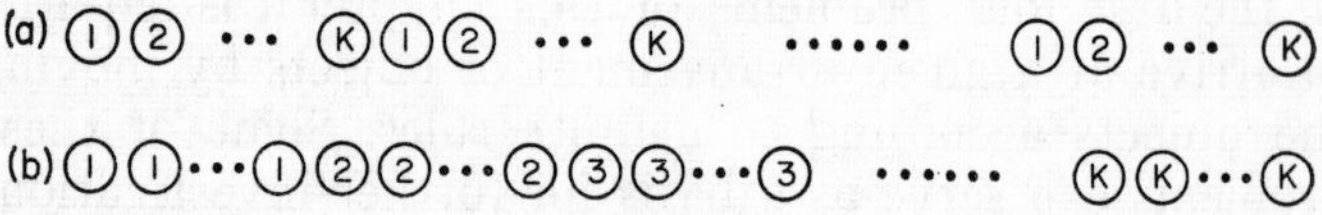

(a) ①② ⋯ Ⓚ①② ⋯ Ⓚ ⋯⋯ ①② ⋯ Ⓚ

(b) ①① ⋯①②② ⋯②③③ ⋯③ ⋯⋯ ⓀⓀ ⋯Ⓚ

Fig. 43.

It can be agreed that each move means the transference of, say, three instead of two neighbouring pieces; and it may be presupposed that the pieces are laid in the new places in their direct order only, or in their reverse order only, or finally, both in the direct and the reverse order. It is also possible to require a different final arrangement of pieces.

Lucas' Problem

The pieces are arranged as in Fig. 44. It is required that the white pieces and the black pieces change places; the white pieces are allowed to move to the right only and the black pieces are allowed to move to the left only, and any piece moves either to a neigh-

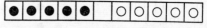

Fig. 44.

bouring empty square, or to an empty square behind the nearest piece of the opposite colour.

128

Objects Changing Places

Having (after an uncomplicated analysis) discarded
moves leading to insoluble positions, we arrive easily
at a solution which can be written down thus
w b b w w w b b b b w w w w w b b b b b w w w w w b b b b
w w w b b w. Here the letters "*w*" and "*b*" denote the
colour of the pieces moved successively.

The Gathering of Coins in Piles

8 coins are arranged in a row. It is required to rearrange the coins in such a way as to obtain four piles
containing 2 coins each.

Here each coin can be joined
only to the coin 2 coins away
from it (either singly or in a
pile). The practically obvious
solution consists of the successive transfer of the fifth
coin on to the second one, the
third one on to the seventh, the fourth on to the first,
and, finally, the sixth on to the eight (see Fig. 45).

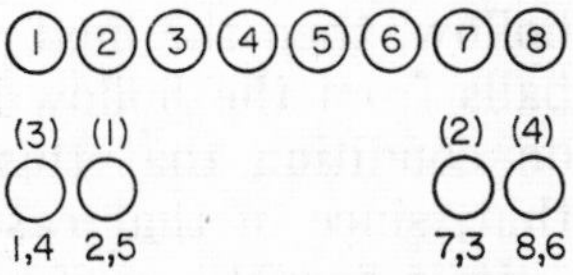

① ② ③ ④ ⑤ ⑥ ⑦ ⑧

(3) (1)   (2) (4)
◯ ◯   ◯ ◯
1,4 2,5   7,3 8,6

Fig. 45.

Try to find the solution of a similar problem for *pn*
coins ($p \geq 4$) which have to be collected in *p* piles, *n*
coins in each, and the transferable coin has to pass
over *n* neighbouring coins (either separate or joined
up — fully or partially — into piles) (see [25]).

Ruma

[34] contains a description of a game of Indian
origin, called "Tschuka-Ruma". Let us demonstrate
it in a somewhat modified form.

$2n + 1$ hollows are distributed in a circle. At the
beginning of the game one hollow is empty (the ruma)
and each of the others contains *n* balls (in Fig. 46 $n = 2$).

The aim of the game is to gather all the balls in the
ruma.

The following action is called a *move*: all balls situated
in some hollow *A* are distributed one by one in the
neighbouring hollows (movement takes place clockwise). If the number of balls to be distributed exceeds

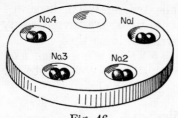

Fig. 46.

$2n$, one ball gets into the hollow A and the remaining balls are again distributed one by one in the neighbouring hollows.

The first move can be made from any hollow. If, after some move (including the first one), the last of the balls gets into the ruma, the next move can be made from any hollow, except the ruma, and if it gets into some other hollow, then the next move should be to distribute the balls from the hollow into which the last ball was put, on condition that this hollow was not empty before that, since in that case the game is regarded as lost.

It is easy to verify that the first move (for $n = 2$) should be made from the hollow No. 3, since in every other case, the second move will cause us to get the last ball into an empty hollow and to lose the game. By trial and error, we can find out that moves made successively from hollows numbered 3, 4, 2, 3, 4, 1, 4, 2, 3, 4, lead us to our goal.

When $n = 3$, the greatest number of balls which can be transferred into the ruma is evidently fifteen. For $n = 2$ there are no less than nine different solutions of the problem.

It would be interesting to construct a theory of the game, or at least to investigate a series of particular cases, in each of which it is required either to find the way (or several ways) of reaching the goal, or to determine the greatest number of balls which can be transferred into the ruma.

It is possible to think of various modifications of the game, by placing, say, s balls ($s \neq n$) in each hollow, by changing the rules of the moves etc.

Objects Changing Places

The Repeated Performance of the Same Operation

There exist a number of card-games based on the fact that by repeatedly disturbing an arrangement of certain objects in accordance with some definite law, we arrive in the end at their initial arrangement.

The explanation of this phenomenon, so puzzling to the uninitiated, is based on simple properties of permutations.

If we have n numbered objects $a_1, a_2, \ldots, a_n$, the transition from their initial arrangement to the arrangement $a\,\alpha_1, a\,\alpha_2, \ldots, a\,\alpha_n$ (where $\alpha_1, \alpha_2, \ldots, \alpha_n$ is some arrangement of numbers $1, 2, \ldots, n$) can be characterized by the permutation $A = \begin{pmatrix} 1 & 2 & 3 & \ldots & n-1 & n \\ \alpha_1 & \alpha_2 & \alpha_3 & \ldots & \alpha_{n-1} & \alpha_r \end{pmatrix}$ showing what number (α_i) should be substituted for the number (i) in the upper row of the permutation A.

The elements $a_1, a_2, \ldots, a_n$ are often denoted by numbers from 1 to n for convenience.

Suppose, for example, the permutation $A =$
$= \begin{pmatrix} 1 & 2 & 3 & 4 & 5 & 6 & 7 & 8 \\ 5 & 8 & 6 & 3 & 7 & 1 & 4 & 2 \end{pmatrix}$ is applied first to the arrangement 1 2 3 4 5 6 7 8 and then to the arrangement obtained in this way, etc. This gives us

$$
\left.
\begin{array}{l}
1\,2\,3\,4\,5\,6\,7\,8 \xrightarrow{A} 5\,8\,6\,3\,7\,1\,4\,2 \xrightarrow{A} 7\,2\,1\,6\,4\,5\,3\,8 \xrightarrow{A} 4\,8\,5\,1\,3\,7\,6\,2 \to \\
\quad\ \text{(I)} \qquad\qquad\quad \text{(II)} \qquad\qquad\quad \text{(III)} \qquad\qquad\quad \text{(IV)} \\
\xrightarrow{A} 3\,2\,7\,5\,6\,4\,1\,8 \xrightarrow{A} 6\,8\,4\,7\,1\,3\,5\,2 \xrightarrow{A} 1\,2\,3\,4\,5\,6\,7\,8 \\
\qquad \text{(V)} \qquad\qquad\quad\ \ \text{(VI)} \qquad\qquad\qquad \text{(I)}
\end{array}
\right\} \quad (1)
$$

We shall give the name of *a product* of two permutations C and D to the permutation which is equivalent to the permutation C and D carried out one after the other (first C and then D).

For example, if $C = \begin{pmatrix} 1 & 2 & 3 & 4 & 5 \\ 2 & 5 & 4 & 1 & 3 \end{pmatrix}$ and $D = \begin{pmatrix} 1 & 2 & 3 & 4 & 5 \\ 5 & 4 & 3 & 2 & 1 \end{pmatrix}$ then $CD = \begin{pmatrix} 1 & 2 & 3 & 4 & 5 \\ 4 & 1 & 2 & 5 & 3 \end{pmatrix}$. Indeed, in the permutation C,

1 is changed to 2, and in the permutation D, 2 is changed to 4, therefore in the permutation CD, 1 becomes 4, etc.

Verify that $DC = \begin{pmatrix} 1 & 2 & 3 & 4 & 5 \\ 3 & 1 & 4 & 5 & 2 \end{pmatrix}$, i. e. $DC \neq CD$.

It is easy to see, that $A^2 = \begin{pmatrix} 1 & 2 & 3 & 4 & 5 & 6 & 7 & 8 \\ 7 & 2 & 1 & 6 & 4 & 5 & 3 & 8 \end{pmatrix}$ is the per-

mutation which changes the permutation I in the scheme (1) into the permutation III, also the permutation II into the permutation IV, etc.

Since the application of permutation A 6 times changes I to I, therefore $A^6 = \begin{pmatrix} 1 & 2 & 3 & 4 & 5 & 6 & 7 & 8 \\ 1 & 2 & 3 & 4 & 5 & 6 & 7 & 8 \end{pmatrix}$,

i. e., all elements remain in their places. A permutation of this kind is called the identity, and is denoted by the letter E.

The smallest natural number s, for which $B^s = E$, is called the *order* of the permutation B. Therefore the order of the permutation A is 6.

For the speedy determination of the order of any permutation (especially when the number of elements being permuted is great) it is convenient to break it down into "independent cycles".

For example, it is easy to notice, that in the permutation A the element 1 becomes 5, the element 5 becomes 7, the element 7 becomes 4, the element 4 becomes 3, the element 3 becomes 6, the element 6 becomes 1, (the cycle is closed).

In other words, the elements 1, 5, 7, 4, 3, 6 replace one another in a cyclic order. This can be characterised by a "cyclic permutation" $\begin{pmatrix} 1 & 5 & 7 & 4 & 3 & 6 \\ 5 & 7 & 4 & 3 & 6 & 1 \end{pmatrix}$, which is

often written down in a one-line form: (1 5 7 4 3 6) or (7 4 3 6 1 5) etc., where it must be remembered, that each element written down is replaced by the next one, and the last element is replaced by the first one (therefore, it is of no consequence which element is the first one to be written down in a cyclic permutation as long as the sequence of the elements is preserved).

In addition, in the permutation A, the element 2 is replaced by 8 and 8 by 2, which gives the cycle (2 8); this is also called the *transposition* of the elements 2 and 8.

Cycles (1 5 7 4 3 8) and (2 8) are called *independent*, since they do not possess any elements in common. Thus A equals the product of two cyclic permutations $A = (1\ 5\ 7\ 4\ 3\ 6)\ (2\ 8)$.

Verify, for example, that $\mathrm{B} \equiv \begin{pmatrix} 1\ 2\ 3\ 4\ 5 & 6\ 7\ 8\ 9\ 10 \\ 9\ 2\ 4\ 1\ 8 & 10\ 5\ 7\ 3\ \ 6 \end{pmatrix} \equiv$

$\equiv (1\ 9\ 3\ 4)\ (2)\ (5\ 8\ 7)\ (6\ 10)$ (here even a single element cycle is to be found).

It is easy to prove[52] that the order of a permutation is equal to the lowest common multiple of the orders of the independent cycles into which the permutation can be broken up. Thus for example, the order of the permutation A is 6 (the lowest common multiple of the numbers 6 and 2) and the order of the permutation B is 12 (the lowest common multiple of 4, 1, 3, 2).

In the scheme (1) it should be noted, that in each of the transitions I→II, II→III, III→IV etc., which are given by the one permutation A, the element from the fifth place is transferred to the 1st place, that from the eight place to the 2nd place, that from the 6th to the 3rd, that from the 3rd to the 4th, that from the 7th to the 5th, that from the 1st to the 6th, that from the 4th to the 7th, that from the second place to the 8th place.

It can be proved that this circumstance is no accident. In addition, the converse also takes place, i. e. if in a series of transitions from one permutation to the second, from the second one to the third, from the third to the fourth, etc., all transitions are carried out according to the same rule — *in the sense of changing the order of places* occupied by the elements — then all these transitions can be characterized by one and the same permutation M, showing which element is substituted for any specific element[53].

In having established the order of the permutation M, we have at the same time found how many times the mutual arrangement of elements has to be changed by the method which is of interest to us, in order to arrive once again at their initial arrangement.

An interesting illustration in connection with the problem discussed above is provided by

Monge's Shuffle

We shall clarify the alteration in the mutual arrangement of objects, known as *Monge's shuffle,* with the aid of the following example: suppose $2n$ pupils are arranged in a row (Fig. 47a) and then they form a double row by means of making every second pupil step behind the pupil beside him (Fig. 47b). After that the second row, headed by the former extreme left-flank pupil carries out an "outflanking manoeuvre" (in accordance with the diagram in Fig. 47c) and finds itself at the right end, while the left-flank pupil is now the extreme right-flank pupil (in Fig. 47 $n = 5$ and the pupils are facing us).

Given a pack containing $2n$ cards, Monge's shuffle is carried out as follows; holding the pack face down in the left hand, we pay out into the right hand each card from the top of the pack in turn, but putting it alternately on top and underneath the cards accumulating in the right hand. Evidently, this operation is characterized by the permutation:

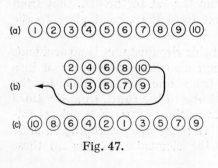

(a) ① ② ③ ④ ⑤ ⑥ ⑦ ⑧ ⑨ ⑩

(b) ② ④ ⑥ ⑧ ⑩
    ① ③ ⑤ ⑦ ⑨

(c) ⑩ ⑧ ⑥ ④ ② ① ③ ⑤ ⑦ ⑨

Fig. 47.

$$M = \begin{pmatrix} 1 & 2 & \dots n-1 \; n \; n+1 \; n+2 \dots 2n-1 & 2n \\ 2n\,2n-2 \dots & 4 \; 2 \; 1 & 3 & \dots 2n-3 \; 2n-1 \end{pmatrix}.$$

For $n = 1, 2, 3, 4, 5, 6, 7, 8, 9, 10, 11, 12, 13, \dots$ the order $M = s = 2, 3, 6, 4, 6, 10, 14, 5, 18, 10, 12, 21, 26 \dots$

We suggest that the reader verifies this table, by calculating the corresponding values of s for various values of n.

For example:

for $n=8$ $M = \begin{pmatrix} 1 & 2 & 3 & 4\,5\,6\,7\,8\,9\,10\,11\,12\,13\,14\,15\,16 \\ 16 & 14 & 12 & 10\,8\,6\,4\,2\,1 \; 3 \; 5 \; 7 \; 9\,11\,13\,15 \end{pmatrix}.$

$M = (1 \; 16 \; 15 \; 13 \; 9) \; (2 \; 14 \; 11 \; 5 \; 8) \; (3 \; 12 \; 7 \; 4 \; 10) \; (6)$ i.e. $s = 5$.

Try to verify (with cards or with any other numbered objects), that the repetition of Monge's shuffle s times leads to the initial arrangement of objects.

It is possible to have a row of $2n$ sportsmen and to carry out s regroupings according to the scheme shown in Fig. 47. The result is the re-establishment of the original order.

There exists a THEOREM ([25], pp. 32–36) which states: *The order of the permutation M is equal to the smallest root of the congruence*

$$2^z \equiv -1 \; (\mathrm{mod}\, 4n + 1),$$

and in the case when this congruence has no solutions, it equals the smallest root of the congruence $2^z \equiv 1$ (*mod* $4n + 1$).

Verify this theorem for various values n. For instance, for $n = 8$, $4n + 1 = 33$: on calculating various powers of 2 we obtain $2^5 \equiv -1$ (mod 33), i.e. for $n = 8$, $s = 5$.

If $n = 5$, $4n + 1 = 21$. For various powers of 2, we have: $2^4 \equiv -5$ (mod 21), $2^5 \equiv -10$ (mod 21), $2^6 \equiv -20 \equiv 1$ (mod 21). Therefore, for $n = 5$, the order of the permutation M equals 6.

§ 23. The Simplest Methods of Constructing Pleasing Patterns

There is scarcely anyone who does not admire the marvellous shapes of snowflakes, the delightful designs of lace created by great master-craftsmen, the complicated drawings on carpets and textiles and patterns on floors inlaid with ceramic tiles of different shapes and colours.

But the creation of pleasing geometric patterns is accessible to everyone, who has sufficient patience.

This and the next few chapters are devoted to geometric pastimes of varying degree of difficulty, whose goal is the obtaining of pleasing patterns, borders, curves, etc. We begin with the simplest ones.

Patterns on squared paper

Anyone can sketch, with little difficulty, various fanciful figures on "squared" paper; here it is not necessary to move along the sides of the little squares only, but diagonals of the little squares and of rectangles to be found on the paper can be drawn in also (Fig. 48).

Similar constructions can be carried out also on paper

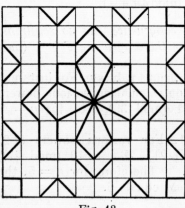

Fig. 48.

with triangular cells, which can be easily made from paper with simple lines ruled on it; let two arbitrary points, A and B (Fig. 49) be situated on one of the lines of the ruled paper; we find a point C, such, that $AC = BC = AB$. Having cut the paper along AC, we obtain a homemade ruler, with intervals of length a marked along its side. With its help we mark off along one of the lines a new sheet of paper the segment $MN = na$ (in Fig. 49, $n = 5$) and, having subdivided it simultaneously into segments of length a, we construct an equilateral triangle MNL, and through the points of subdivision of its sides we draw lines parallel to the sides ML and NL.

In the construction of a border, it may be permitted to draw, in the triangular network so obtained, the medians of the triangles, which is equivalent to the using of a network with triangles having a side of $\frac{a}{2}$ (see Fig. 50).

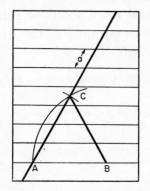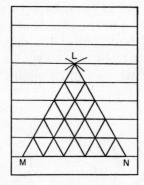

Fig. 49.

Taking as a basis a square or a triangular network, try to arrange a competition for the best pattern, taking into account the originality of the drawing, the accuracy of execution and the quality of colouring (the size of the drawing must be agreed upon, for

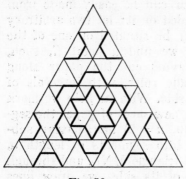

Fig. 50.

example, a rectangle of 12 × 20 squares, a big square of 10 × 10 squares, a hexagon of side 6*a*).

Using Compasses and a Ruler

By making use of compasses and a ruler it is possible to construct the most varied figures. It is convenient to make use of a circumference subdivided into *n* equal parts (for *n* = 3, 4, 5, 6, 8, 10 etc., this is easily done, and for *n* = 7, 9, 11 etc. and even for *n* = 5, 10, a protractor can be used). In order to save labour, it is possible, having first checked the construction very carefully, to pierce the centre and the vertices of a regular polygon with a sharp needle, and so carry the marks through onto several clean sheets of paper.

By drawing chords of equal length through the points of subdivision of the circumference, we obtain various regular starlike polygons.

If we now draw circles (or arcs) of various radii and

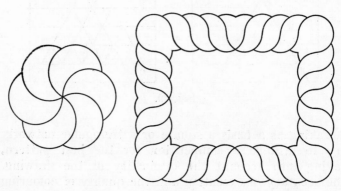

Fig. 51.

centres at the vertices of regular polygons and at the points of intersection of circles, and if we then join the different pairs of points by straight lines, we can obtain an infinite variety of figures, whose attractiveness is enhanced by successful colouring.

In addition to a circle, other figures can be used as a base — rectangle, triangle, etc. (see Fig. 51).

Symmetrical Figures

Two points, A and A' are called *symmetrical* with respect to the straight line l, if A and A' are equidistant from and on opposite sides of l, and $A A'$ l.

A plane figure is called symmetrical with respect to a straight line l, if for any point B on it there can be

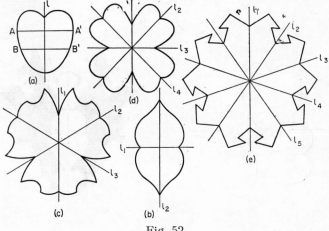

Fig. 52.

found a point B' (also belonging to the figure), symmetrical to the point B with respect to the line l. The straight-line l is called an *axis of symmetry*.

Figure 52 depicts figures possessing 1, 2, 3, 4 and 5 axes of symmetry respectively.

Figures with several axes of symmetry can be easily cut out of paper: a sheet of thin paper is folded double and then folded again n times in the form of a "sector" with a central angle of $\dfrac{180°}{n}$: on cutting the sector along some curve and unfolding the paper, we obtain a figure possessing n axes of symmetry. If coloured paper is used, we obtain lacy doilies in various colours with fancy patterns.

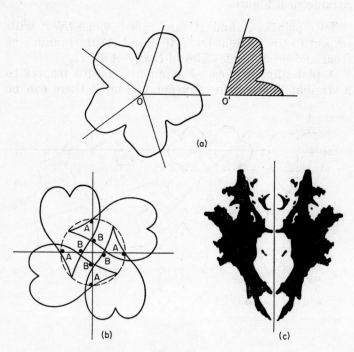

Fig. 53.

We shall call the point 0 *the centre of symmetry of the nth order* of the given figure, if on rotating the figure about the point 0 by an angle of $\dfrac{360°}{n}$, it coincides with its original position. For example, in Fig. 52, the shapes *b*, *c*, *d*, *e* possess centres of symmetry of the 2nd, 3rd, 4th and 5th order respectively.

For the construction of figures possessing central symmetry of the nth order, a cardboard template in the form of a sector with a curved edge and an angle of $\frac{360°}{n}$. is useful.

Let us draw n rays from the point 0, forming angles of $\frac{360°}{n}$. By placing the cardboard template between each pair of neighbouring rays in turn, (so that the point 0' coincides with 0) and outlining in pencil the outer edge of the sector, we obtain a symmetrical figure similar to the one depicted in Fig. 53a.

One can also use a template of any desired shape and superimpose it on each ray in turn in one particular way, for example, so that two of its points, A and B are situated on each ray at the same distance from 0 in each case (Fig. 53b).

We shall mention one more method of constructing symmetrical figures — the method of folding a sheet of paper containing an ink blot and pressing the two halves together. Some very fanciful figures with one axis of symmetry are often obtained in this way (Fig. 53c).

§ 24. Regular Polygons from Rhombi

It can be seen from Fig. 54a, that if we take a sep-
tangular star made up of 7 rhombi of side a and acute
angle $\alpha = \frac{2\pi}{7}$, and place a second layer of rhombi of
angle $\beta = 2\alpha = \frac{4\pi}{7}$, between its arms, then place yet
another 7 rhombi of angle $\gamma = 3\alpha = \frac{6\pi}{7}$ in the gaps
of the new star, we obtain a regular 14-sided polygon
of side a.

Having taken an octangular star (Fig. 54b) consisting
of 8 rhombi of angle $\alpha' = \frac{2\pi}{8}$, we have a second layer
of squares $\left(\beta' = 2\alpha = \frac{\pi}{2} \right)$ and a third layer of rhombi
exactly like those in the central star $(\gamma' = 3\alpha' = \pi - \alpha')$.
All 3 layers of rhombi form a regular octagon of side $2a$.

Attempt to prove that, if $\alpha = \frac{2\pi}{m}$ and m is odd
$(m \leq 3)$, then $\frac{m-1}{2}$ layers of rhombi form a regular
$2m$ sided polygon of side a, and if m is even $(m \geqslant 4)$
then $\frac{m-1}{2}$ layers of rhombi form a regular m-sided
polygon of side $2a$[54]. Since each layer has m rhombi,
it follows that for m odd, any regular $2m$-sided polygon
of side b can be subdivided either into $2m (m - 1)$
rhombi of side $\frac{b}{2}$, or into $\frac{m(m-1)}{2}$ rhombi of side b
(see, e. g. Fig. 54b, c).

Note that the subdivision of a regular $2m$-sided
polygon of side $2a$ into rhombi of side a may be obtained
by rotating the smaller regular $2m$-sided polygon of

side a about one of its vertices consecutively through angles $\frac{\pi}{m}$, $\frac{2\pi}{m}$, $\frac{3\pi}{m}$, ..., $(2m - 1)\frac{\pi}{m}$ (see $ABCDEFGH$ and $ABCDEFGHJK$ in Fig. 54 c, d). The smaller $2m$-sided polygon which is to be rotated, consists of the same kinds of rhombi as the large one, but it has only a quarter of the number of rhombi of each kind than has the large $2m$-sided polygon.

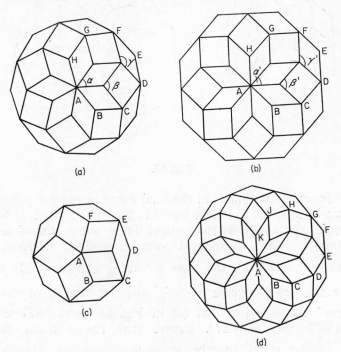

(a)   (b)

(c)   (d)

Fig. 54.

The subdivision of a $2m$-polygon of side a into rhombi of side a (m being odd!) can be obtained by rotating the equilateral $(m + 1)$-angled figure of side a, in which two opposite angles equal $\frac{\pi}{m}(m - 1)$ and each of the remaining $m - 1$ angles equals $\frac{\pi}{m}(m - 2)$, about a vertex with angle $\frac{\pi}{m}$ $(m-1)$, by $\frac{2\pi}{m}$, $2\frac{2\pi}{m}$, ...$(m-$ $-1)$ $\frac{2\pi}{m}$ radian (see, e.g. in the fourteen-sided polygon (a) the octagon $ABCDEFGH$ and in the decagon (a) the hexagon $ABCDEF$).

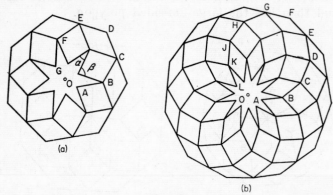

Fig. 55.

In order to obtain, in the final count, a regular polygon of m sides, each side equal to $2a$, m being odd, it is required to take as the central figure not a star of m rhombi, but a regular star-shaped m-sided polygon, whose angle α at the apex equals $\frac{\pi}{m}$ and the angle β at the "gap" equals $3\alpha = \frac{3\pi}{m}$ and the "depth of the gap" $AB = a$ [(a) and (b) in Fig. 55 correspond to $m = 7$ and $m = 11$]. Prove, that the addition of $\frac{m-3}{2}$ layers of rhombi to such a polygon always leads to a regular m-sided polygon of side $2a$[55]. Here the whole subdivision of the m-sided polygon into rhombi can be carried out with the help of m small "open" regular polygons [like $ABCDEFG$ in Fig. (a) or $ABCDEFGHJKL$ in Fig. (b)] which change one into the next on rotating about the centre of the star by the angle $\frac{2\pi}{m}$.

144

§ 25. The Construction of Figures from Given Parts

Mosaic

In the game called "mosaic" various figures are made up of a definite set of coloured tiles. The sets usually consist of squares, rhombi and right-angled triangles with an acute angle of 45° (Fig. 56).

In the simplest case, all tiles are placed according to drawings enclosed with the game. But it is possible to try to find original figures, without reference to these drawings and without restricting oneself as to the shape of the figure, using all the tiles given or some part of them.

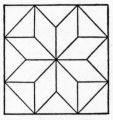

Fig. 56.

Perhaps the reader might invent new interesting variants of the game with sets of tiles containing various equilateral triangles, parallelograms, regular polygons, etc. The variants invented may be realised on thick cardboard by sticking on bits of coloured paper.

If we use, instead of cardboard or wooden tiles, thin coloured laminae made of glass or transparent plastic, and permit overlapping, then even a two-colour mosaic may yield considerable variety and colourfulness of design when viewed against the light.

Figures out of Pieces of a Square

One useful and enjoyable pastime is the construction of figures out of the 7 portions of a square obtained

145

when it is cut up according to Fig. 57a. In the construction all seven pieces must be used up and no amount of overlapping is allowed.

Figure 58 shows symmetrical figures, borrowed from the book by V. I. Obreimov [19]. Try to construct these figures out of the parts of the square given in Fig. 57a.

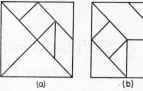

Fig. 57.

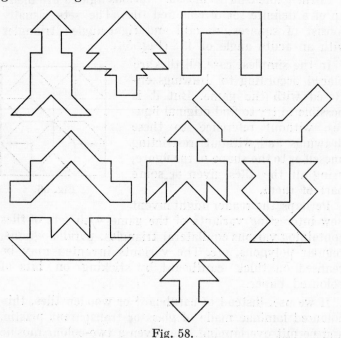

Fig. 58.

These parts can be fitted together to form many other figures (for example, pictures of various objects, animals, etc.

A less widespread variant of this game (see [23] part 1, p. 209) is the construction of figures out of portions of the square depicted in Fig. 57b.

Constructions from Given Parts

Hold a competition in inventing figures that can be constructed out of these portions. The winner is the person, who is quickest at constructing figures proposed by his opponent.

Could you manage to subdivide a square into 7 portions in a different way, so that various symmetrical figures may be constructed out of them?

If desired, the game may be made more complicated by inventing figures to be fitted together, say, out of six or five pieces of the square and without indicating which pieces are to be used.

Rectangles Out of Squares

In the last 20 years there appeared in certain mathematical magazines a number of articles dealing with the question of constructing rectangles out of squares, no two of which are alike.

It turns out that it is impossible to construct a rectangle out of n different squares, if $n < 9$. When $n = 9$, the problem has two solutions; it is possible to construct a rectangle out of squares whose sides are in the ratio 1:4:7:8:9:10:14:15:18 (Fig. 59) and also out of squares, whose sides are in the ratio 2:5:7:9:16:25:28:33:36. Construct[56] a rectangle out of 10 squares whose sides are in the ratio 3:11:12:23:34:35:38:41:44:45, and one out of 13 squares with sides being in the ratio 1:4:5:9:14:19:33:52:56:69:70:71:72 (it is useful to solve first the question of the ratio of sides of each of the rectangles under construction).

The smallest number of various squares from which a large square can be constructed is 26. Construct a large square, taking into account that the sides of the squares making it up are to each other as 1:11:41:42: 43:44:85:168:172:183:194:205:209:5:7:20:27:34:61:95: 108:113:118:123:136:231, and that the first 13 squares form a rectangle with sides in the ratio 608:377, and the last 13 squares form a rectangle with sides in the ratio 608:231 (57).

In the book by B. A. Kordemskii and N. V. Rusalev [14] the connection is noted between the problem of constructing rectangles out of squares and the problem of determining the distribution of currents in closed circuits composed of several conductors.

It is interesting to note that the problem of constructing a right-angled parallelepiped out of a finite number of cubes, no two of which are alike, is insoluble.

Indeed, suppose the problem has a solution and v is the volume of the smallest cube. Since in any rectangle constructed out of squares, no two of which are alike, the smallest square cannot be next to the side of the rectangle (prove this[58]), therefore the smallest of the cubes (K_1) next to the lower base of the parallelepiped is surrounded by cubes of greater dimensions; the smallest of the cubes (K_2), at the bottom of the "well" so formed, is also surrounded by larger cubes, which form a well with a smaller cross-section, and so on.

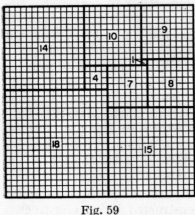

Fig. 59

Among the cubes K_1, K_2, K_3,..., there will appear, finally, cubes whose volumes are less than v_1 and this contradicts the initial supposition.

Solve also the following two problems: (1) Subdivide a cube into n cubes (among which there may be identical ones) for $n = 34$ and for $n = 50$[59]. Try to establish for what values of n this problem is insoluble. (2) Prove[60] that for $n \neq 2, 3, 5$ a square can be subdivided into n squares (among which there may be identical ones).

§ 26. The Construction of Parquets

One interesting geometrical pastime is the construction of parquets — this is the short description of the activity of covering a plane with figures of a certain shape or several given shapes in a certain regular sequence.

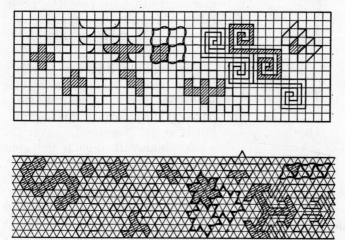

Fig. 60.

Examples of the simplest kind of parquets are: ordinary squared paper and a plane covered by identical regular triangles. If the separate units are combined into complexes by some method or other, a wide variety of different "derived" parquets may be obtained. The variety may be increased, if some units are increased in size at the expense of others, by subdividing the others by curves or various straight lines. Figure 60 shows several examples of derived parquets, which extend indefinitely (we suggest that the reader verifies that).

Derived parquets can also be constructed out of parquets composed of regular polygons with various numbers of sides.

In order to investigate the question of filling a plane with regular polygons it is necessary, firstly, to take into account all possible kinds of nodal points, i. e. to find various combinations of regular polygons, whose vertices concur at a point, and which cover, without overlapping, the vicinity of that point, and verify in each case the possibility of infinite continuation of a parquet with nodal points of the type found.

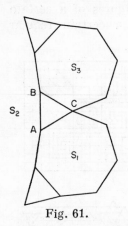

Fig. 61.

For example, it is easy to verify that a regular triangle, a regular septagon, and a regular 42-sided polygon may form a nodal point, but it is not possible to cover the whole plane with regular polygons forming nodes of the type (3, 7, 42) only. Indeed, if the triangle ABC, the septagon S_1, and the 42-sided polygon S_2 concur at the vertex a (Fig. 61), then the vertex B should be the point at which the septagon S_1 should join the triangle ABC and the 42-sided polygon, but then the point C cannot be a nodal point of the type (3, 7, 42).

Prove that, although nodes of the type (5, 5, 10 do exist, it is not possible to fill a plane with regular pentagons and decagons([61]).

In searching for various types of nodal points, it should be noted that the order k of a nodal point (i. e. the number of polygons concurring there) cannot

Construction of Parquets

exceed 6. In addition, if regular n_1, n_2, ..., n_k-sided polygons concur at a node of the kth order, it is easy to prove[62] that

$$k-2\left(\frac{1}{n_1} + \frac{1}{n_2} + \ldots + \frac{1}{n_k}\right) = 2.$$

For example, for $k = 3$, we have the relation $\frac{1}{n_1} + \frac{1}{n_2} + \frac{1}{n_3} = \frac{1}{2}$, which is satisfied by the following nodes (8, 8, 4), (12, 12, 3), (10, 5, 5), (12, 6, 4) (6, 6, 6) (3, 7, 42), etc.

For $k = 4$, we have $\frac{1}{n_1} + \frac{1}{n_2} + \frac{1}{n_3} + \frac{1}{n_4} = 1$; this condition is satisfied by nodes: (6, 3, 4, 4), (6, 3, 6, 3),etc.

It is impossible to exhaust all parquets. Indeed, taking, say, a triangular network and joining up some of the triangles into hexagons in various ways, we can obtain an infinite variety of parquets constructed out of triangles and hexagons (Fig. 62).

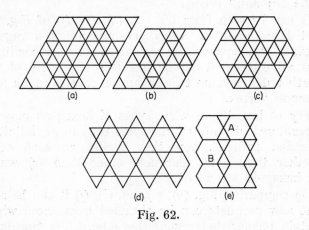

Fig. 62.

Therefore it is advisable to lay down quite rigid limiting conditions in constructing parquets. For example, there can be a requirement that the sets of polygons surrounding each node be the same for each node. This requirement is not fulfilled in parquets

11•   151 (*a*), (*b*) and (*c*), since here some nodes are of the fifth order and some of the sixth order, but it is fulfilled for parquets (*d*) and (*e*).

If an additional condition is imposed, namely, that the mutual disposition of the polygons adjoining some node of the parquet should be the same as for all the other nodes, then parquet (*d*) must fail, since at the node *A* triangles and hexagons alternate, but at the node *B* both hexagons are side by side.

It can be required, that all identical polygons of the parquet are single-type in the sense that any two polygons of the same name, together with their adjacent polygons can be made to coincide with each other fully [see (*a*), (*b*), (*c*), (*d*) in Fig. 63].

Or it may be permitted to make use of polygons of two different types [for example, in the parquet (*e*) all squares and all hexagons are single-type, but not the triangles, since some triangles have 3 squares adjoining them, and others have 2 squares and a triangle adjoining them].

It can be seen from (*c*), (*g*), (*h*), (*i*), (*j*) of Fig. 63, that there are five ways of surrounding a regular hexagon with squares and regular triangles on condition that all 6 vertices of the hexagon are nodes of the fourth order, and no two hexagons have so much as a common vertex.

Try to investigate what types of hexagons may be encountered at one and the same time in an infinitely extended parquet, how they border on each other whether there exist parquets other than (*c*) where the hexagons are all of one type.

On comparing Fig. (*a*), (*b*), (*c*), (*e*), (*f*) it can be seen that new parquets can be obtained from those which contain regular dodecagons (by subdividing the dodecagons into constituent parts).

By joining up each four neighbouring triangles and the corresponding two neighbouring squares in the parquet (*f*) into complexes, we obtain a parquet, all of whose units and all of whose nodes are single-type,

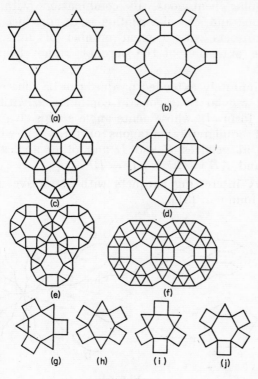

Fig. 63.

but irregular units make their appearance (rectangles made up to two squares).

Let us note an interesting property of equilateral hexagons, two of whose opposite angles are $\dfrac{360°}{n}$ each (n is some natural number) and the remaining angles are equal. Using tiles of this kind it is possible to construct, in addition to the parquet (a) of Fig. 64 with parallel units, the parquet (b), where the units fill n sectors of form AOB, concurring at centre 0 (in the drawing, $n = 7$) and n similar sectors, BCD, $AEF, \ldots$, whose vertices do not quite reach the centre.

153

Regular pentagons, in combination with regular decagons and star-like pentagons, can produce interesting parquets with a great number of units, but an infinite extension of the parquet cannot be achieved here.

An infinitely extended parquet can be constructed[63] out of regular pentagons in combination with rhombi of the form (*d*), whose acute angle equals 36°, and also out of equilateral pentagons of form (*c*), with right angles at vertices *B* and *D* and other angles of 120° each and $AB = BC = CD = DE$.

Many interesting parquets with attractive units are to be found in [31].

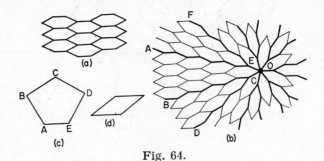

Fig. 64.

The construction of parquets and, in particular, their investigation is accessible to persons with a certain amount of mathematical preparation. But the colouring of a parquet that has been constructed already, the joining up of its units into complexes and even the changing of the form of the units of a prepared parquet, does not require any mathematical preparation.

Two-colour Parquets

The book by V. I. Obreimov [19] contains the description of square parquets, which are composed of square tiles divided diagonally into two equal triangles (a white one and a black one).

Taking a complex of four little squares, for instance
ABCD in Fig. 65, it is possible to build on to it (on
the right and underneath), three other complexes, in
such a way that a 16-square symmetrical parquet is
obtained, possessing a horizontal (*AE*) and a vertical
(*CF*) axis of symmetry.

Complexes of 9, 16, 25, etc., squares may be taken
instead of one of 4.

Since it is possible to construct 4^{n^2} various complexes
out of n^2 squares therefore it is practically impossible
to exhaust the number of symmetrical parquets built
up of $4n^2$ squares, for $n > 2$.

If, in a symmetrical parquet, any four squares, symmetrical with respect to both axes of symmetry of the
large square, for instance the squares α, β, γ, δ, in
Fig. 65, were to change places without losing their
orientation, in the horizontal direction (i. e. if we carry
out transpositions (α, γ) and (β, δ) or in the vertical
direction (transpositions (α, β) and (γ, δ)), we again
obtain a symmetrical parquet.

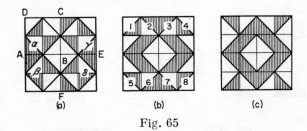

Fig. 65

Thus, horizontal transpositions (α, γ) and (β, δ) transform parquet (*a*) into parquet (*b*) and the latter is
transformed by the vertical transpositions (1, 5), (2, 6),
(3, 7) and (4, 8) into the parquet (*c*).

We shall call two parquets *related* or *belonging to
the same class*, if we can pass from the one to the other
by means of one or several transpositions of parallel
rows of squares, symmetrical with respect to the
horizontal or vertical axes of symmetry of the parquet.

For example, the parquets (*b*) and (*c*) in Fig. 65 are related since they are obtained from each other by transposing the top and the bottom horizontal rows.

The reader should determine, for $n = 2, 3, 4, \ldots$, the number of the many various classes into which sets of symmetrical multisquare parquets, consisting of $4n^2$ squares can be divided.

It is also possible to construct hexagonal parquets from triangles painted in two or three colours [see Fig. 66*a*, *b*].

On constructing a large equilateral triangle *ABC* (Fig. 66*c*, where $n = 2$) or *A'B'C'* (Fig. 66*d*, where $n = 3$) out of 4 (9, 16, and, in general, n^2) triangles of form (*a*), and reflecting it in the side *BC* (*B'C'* respectively) and reflecting the rhombus *ABCD* (*A'B' C'D'*) thus obtained in *AB* and *BD* (in *A'B'* and *B'D'*) we obtain a regular hexagon with three axes of symmetry [see (*e*) and (*f*)] consisting of $6n^2$ small two-coloured triangles.

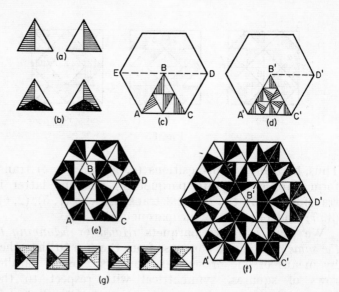

Fig. 66.

If you carry out the reflection in the reverse order (first the triangles *ABC* in the side *AB*, and then the rhombus obtained from sides *BE* and *BC*) the same hexagon is obtained([64]).

The most varied square parquets with the most fantastic designs can be laid out from $4n^2$ tiles of four colours; since 4 triangles with a common vertex at the centre of the square can be painted in 6 substantially different ways [Fig. 66g], therefore we can agree either to use up all 6 sorts of squares, or to introduce some sort of restrictions.

§ 27. Re-cutting of Figures

Two figures (by that we shall understand two plane figures or two bodies) are called *equi-composed* if one of them can be subdivided into parts, which when joined together give the other figure.

We shall call this process the *recutting* of one figure to form the other.

THEOREM I. *If each of the figures A and B is equi-composed with a certain figure C, then the figures A and B are also equi-composed with each other.*

Indeed, if figure C (Fig. 67) is subdivided by continuous lines (planes for solid bodies) into parts which can make up figure A, and by dotted lines into parts forming figure B, then the set of numbered parts, subdividing C by the two kinds of lines is a collection of pieces, out of which the figure A can be put together by joining pieces 1, 2; 3, 4, 5 etc. and the figure B can be put together by joining pieces 1, 3, 12; 2, 4, 6, 9 etc.

The following theorem is self-evident:

THEOREM II. *Any two equi-composed figures are equal in area.*

The converse theorem does not hold, i.e. the equi-composition of two figures does not follow, generally speaking, from their equality in area. Several theorem squoted below refer to special cases, where equality in are a of two figures leads to their equi-composition.

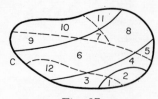

Fig. 67.

158

THEOREM III. *Parallelograms with equal bases and equal altitudes are equi-composed (see Fig. 68a).*

THEOREM IV. *Any two rectangles that are equal in area are equi-composed.*

Using the condition $ab = cd$, show that $AB \parallel CD \parallel LM$ (Fig. 68b); on drawing $EDF \parallel AB$, it is easily seen that $S_1 = S'_1 \colon S_2 = S'_2$; and S_3 and S'_3 are equi-composed).

THEOREM V. *Any two parallelograms, which are equal in area, are equi-composed.*

(On the basis of Theorem III, each of these parallelograms is equi-composed with some rectangle; it remains to apply Theorems I and IV.)

THEOREM VI. *Any triangle and rectangle that are equal in area, are equi-composed.*

(It should be noted that any triangle ABC is equi-composed with some parallelogram $ADFC$, Fig. 68c).

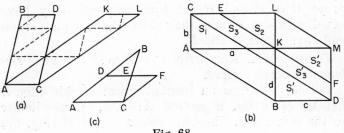

Fig. 68.

THEOREM VII. *Any two polygons, which are equal in area are equi-composed.*

Indeed, having subdivided each polygon (P and Q) into triangles (p_1, p_2, ..., p_m and q_1, q_2, ..., q_n, respectively), it is sufficient to recut all triangles into rectangles (p'_1, p'_2, ..., p'_m and q'_1, q'_2, ..., q'_n) with a common altitude h. The rectangle with the altitude h, composed of p'_1, p'_2, ..., p'_m (it can be composed of q_1, q_2, ..., q_n as well) is equi-composed with P and Q.

As we see, the theorem, converse of the Theorem II, is true for any polygons.

In the last century attempts were made to prove that a theorem, analogous to the Theorem VII is true for polyhedra. But in 1901 Denn, a German mathematician proved that a cube and a tetrahedron that are equal in volume, are not equi-composed (the proof of Denn's Theorem, and many interesting details connected with the equi-composition of figures, can be found in the book by V. G. Boltyanskii [5]).

If we however, limit ourselves to polyhedra of special form, then in some cases their equi-composition follows from their equality in volume.

THEOREM VIII. *Rectangular parallelepipeds, which are equal in volume, are equi-composed.*

Let us denote these rectangular parallelepipeds by P and Q. Let a, b, c be the edges of P and a_1, b_1, c_1, be the edges of Q and let $abc = a_1 b_1 c_1$. We take an auxilliary rectangular parallelepiped, R, with edges a_1, b', c, where $a_1 b' = ab$ and $b'c = b_1 c_1$; it is easy to prove that P and R are equi-composed (they have equal altitude, c, and their bases have equal areas) and that so are Q and R.

THEOREM IX. *Any prism can be recut into some rectangular parallelepiped.*

Indeed, let P be an inclined prism of side edge l. Let us carry out a section $ABCDE$, perpendicular to the side edges. Then the pieces I and II (Fig. 69) make up a right-angled prism P', with height l and base $ABCDE$ (if a plane, perpendicular to side edges and cutting all of them cannot be constructed — perhaps the prism is too broad and short — then the prism can be first subdivided into narrow prisms with a smaller base, and then each of these recut in the way prescribed into a right-angled prism).

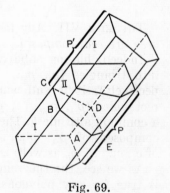

Fig. 69.

Re-cutting of Figures

According to Theorem VII, by subdividing a polygon *ABCDE* into pieces $\alpha_1, \alpha_2, \ldots, \alpha_n$, it can be recut into some rectangle; it follows that a rectangular parallelepiped *P″* of height *l* can be constructed out of the right-angled prisms, with bases $\alpha_1, \alpha_2, \ldots, \alpha_n$ and heights *l*, which make up *P′*.

THEOREM X. *Any two prisms of equal volumes are equi-composed* (this follows from Theorems VIII and IX).

Problems on recutting figures are often accompanied by indicating the number of parts into which they have to be subdivided. For example:

1. By subdividing a rectangle 9 × 16 cm² into two parts, recut it into a square ([65a]).

2. By subdividing a rectangle $a \times b$ cm² into two parts recut it into a rectangle $\dfrac{an}{n+1} \times \dfrac{b(n+1)}{n}$ cm² (*n* is a natural number) ([65b]).

3. By subdividing a rectangular parallelepiped 8 × × 8 × 27 cm³ into 4 parts, recut it into a cube of edge 12 cm (see Fig. 70).
Try to prepare cardboard models of four parts, which can be used both for the construction of a cube of edge 12 cm and for the construction of a parallelepiped 8 × 8 × 18 cm³.

4. By subdividing a rectangular parallelepiped *abc* cm³ into 4 parts, recut it into a parallelepiped $\dfrac{am}{m+1} \times \dfrac{b(m+1)n}{m(n+1)} \times \dfrac{c(n+1)}{n}$ cm³ (*m* and *n* are integers) or into a parallelepiped

$$\frac{amn}{(m+1)(n+1)} \times \frac{b(m+1)}{m} \times \frac{c(n+1)}{n} \text{ cm}^3 \quad ([66]).$$

5. Figure 71 shows how to cut up, each into two parts, a

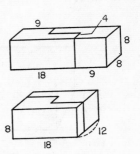
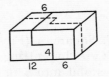

Fig. 70.

161

rug (a) and a piece of check oilcloth (b) in order to make of the pieces obtained, a square rug in the first case and a chessboard in the second case.

Try to think of similar problems in which the constituent parts have three, four, six, points.

6. Triangles ASB, BSC, CSD (Fig. 72) have a common vertex at S and equal bases along the same straight line and next to each other ($AB = BC = CD$).

Prove that each of them is composed of the same set of pieces, numbered 1, 2, 3, 4 ($BK \parallel CL \parallel AS$; $BE \parallel CF \parallel DS$; $BM \parallel CS$; $CN \parallel BS$).

Investigate an analogous problem for n triangles ($n = 4, 5, 6, \ldots$) having a common vertex and equal, adjoining bases in one straight line, supposing that they are subdivided by means of straight lines passing through the vertices adjacent to the bases of the triangles, and which parallel the end sides of all triangles together.

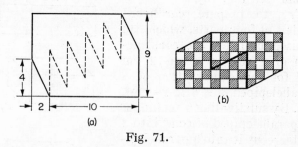

(a)

(b)

Fig. 71.

7. Square A (Fig. 73) is to be recut ([67]) into figures B, C, D by cutting it up into 3 parts in the first two cases and into 4 in the last case.

N o t e: In your mathematical circle you can organise a competition in inventing interesting figures, which can be built up from a definite number of parts of a square, or any given figure (circle, regular hexagon, etc.).

Figure 74 shows how six numbered pieces of a regular pentagon
ABCDE of side *a*, can be put
together to form an equilateral
triangle of side $b = a\sqrt{\left(\dfrac{5}{3.\tan 36°}\right)}$
$\cong a \times 1\cdot993$ (see [39], 1952, No.
2, p. 106). Constructions are as
follows

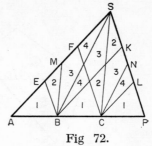

Fig 72.

(1) *DK* is *BD* produced and
  DK = DE, i.e. $\triangle EDK = \triangle BCD$;
(2) *EL = LK*; *MN* ∥ *AB*: *CP = DM*; *BQ = EL*,
  i. e. *EDML = BCPQ* and $\triangle QPD = \triangle LMK =$
  $= \triangle LNE$;
(3) $ET = \dfrac{1b}{2} = ES = TS = TU = EV$, i. e. the
  equilateral triangle *SUV* is equal in area to the
  pengaton *ABCDE*:

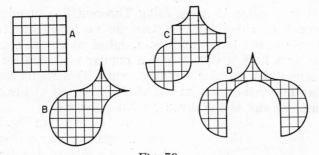

Fig. 73.

(4) *RW* ∥ *UV* (but *RW* and *AB* are not parallel
  therefore $\triangle TUF = \triangle TSR$ and $\triangle EHF =$
  $= \triangle EWS$. In addition, since the quadrilaterals
  ABRW and *NMFH*, whose corresponding sides
  are parallel, are equal in area, it follows that
  they coincide.

In the book by B. A. Kordensky and N. V. Rusalyov
[14] it is shown how to recut a square into a regular
triangle, a regular hexagon and a regular pentagon,

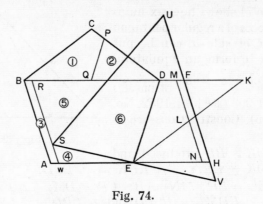

Fig. 74.

by subdividing it into 5, 5, and 6 pieces. Try to recut a regular septagon into a regular triangle or into a square, a regular hexagon into a regular octagon, etc. Here one should strive to achieve the smallest number of pieces of a reasonably large size.

It is possible that by using Theorem I, you might succeed in subdividing a triangle (a hexagon, etc.) into pieces not too small in size, out of which one could construct both a square and a regular pentagon (or a square and a regular septagon, etc.). This would lead to an interesting puzzle: out of the given set of pieces, construct any of the given regular polygons.

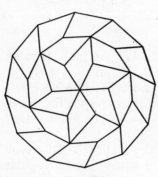

Fig. 75.

It can be seen from Fig. 75 that a regular dodecagon can be subdivided into identical pieces in such a way that out of a double supply of these pieces a regular dodecagon of double the area may be constructed.

Investigate the following problem; what are the values of n and k, for which a regular n-sided polygon can be split up into identical pieces (as few as possible) so that k sets of such pieces can be used to construct a similar figure of area k times as large as the original one.

The reader should try to subdivide a square into n pieces such that on discarding one of them, 2 (or even 3) of the remaining pieces can be rearranged each time into a smaller square.

A similar question can be set for a regular triangle, pentagon, etc.

§. 28. The Construction of Curves

Rosettes

Lovers of attractive geometrical figures can occupy themselves in drawing curves with equations in polar coordinates, of the form $r = a + b \sin\left(\dfrac{m\Phi}{n}\right)$, where a, b, m, n are given numbers.

We select a unit of length e, a pole O and a polar axis Ox (Fig. 76) for polar coordinates. The position of any point M is determined by the polar radius OM and the polar angle φ, formed by the ray OM and the ray Ox. The number r such that $OM = re$ and the numerical value of the angle φ, expressed in degrees or in radian, are called the *polar coordinates of the point M*.

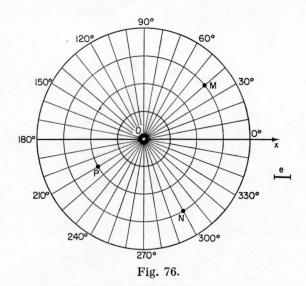

Fig. 76.

Construction of Curves

For any point other than the pole O, it can be taken that $0 \leq \varphi \leq 2\pi$ and $r > 0$. However, in drawing curves, corresponding to equations of the form $r = f(\varphi)$. it is natural to give the variable φ any values (including negative ones, and those exceeding 2π) and r might be either positive or negative.

In order to find the point (φ, r) we draw a line from the point O, to form an angle φ with the axis Ox, and we measure off along it (if $r > 0$), or along its continuation in the opposite direction (if $r < 0$) a segment of length $|r|$ e.

Everything is much simplified if a coordinate network is constructed consisting of concentric circles of radii e, $2e$, $3e$, etc. (with centre at the pole O) and rays, for which $\varphi = 0°, 10°, 20°, \ldots, 340°, 350°$; these rays are of use both for $\varphi < 0°$ and for $\varphi > 360°$; for example, for $\varphi = 740°$ and for $\varphi = -340°$, we arrive at the ray for which $\varphi = 20°$ [see in Fig. 76 points $M(40°, 3)$, $N(120°, -3)$, $P(-1230°, 2)$].

Let us consider the curves:

I. $r = \sin 3\varphi$, II. $r = \dfrac{1}{2} + \sin 3\varphi$,

III. $r = 1 + \sin 3\varphi$, IV. $r = \dfrac{3}{2} + \sin 3\varphi$.

We compile a table

φ	—30°	—20°	—10°	0°	10°	20°	30°
$\sin 3\varphi$	—1	—0·87	—0·5	0	0·5	0·87	1
$\dfrac{1}{2} + \sin 3\varphi$	—0·5	—0·37	0	0·5	1	1·37	1·5
$1 + \sin 3\varphi$	0	0·13	0·5	1	1·5	1·87	2
$\dfrac{3}{2} + \sin 3\varphi$	0·5	0·63	1	1·5	2	2·37	2·5

φ	40°	50°	60°	70°	80°	90°
$\sin 3\varphi$	0·87	0·5	0	—0·5	—0·87	—1
$\dfrac{1}{2} + \sin 3\varphi$	1·37	1	0·5	0	—0·37	—0·5
$1 + \sin 3\varphi$	1·87	1·5	1	0·5	0·13	0
$\dfrac{3}{2} + \sin 3\varphi$	2·37	2	1·5	1	0·63	0·5

$\left(\text{Here 0·87 is the approximate value of } \dfrac{\sqrt{3}}{2}.\right)$

We shall call a portion of a plane, for whose points $\alpha \leq \varphi \leq \beta$, a sector. On joining the points (0°; 0), (10°; 0·5), (20°; 0·87), (30°; 1), (40°; 0·87) (50°; 0·5), (60°, 0) by a smooth curve we obtain a "positive loop" (1) of the curve I, in the sector (0°, 60°) (see Fig. 77a). On continuing the table to $\varphi = 360°$ and joining the points (60°; 0), (70°; —0·5), (80°; —0·87), (90°; —1), (100°; —0·87), (110°; —0·5), (120°, 0) of the drawing, situated in the sector (240°, 300°), we obtain in this sector, a "negative loop" (2).

It is easy to find out that it will be followed by a positive loop (3) in the sector (120°, 180°) , a negative loop $\overline{(4)}$ in the sector (0°, 60°), positive loop (5) in the sector (240°, 300°) and, finally, negative loop $\overline{(6)}$ in sector (120°, 180°).

In the curve I, the loops (1) and $\overline{(4)}$ coincide and so do $\overline{(2)}$ and (5), (3) and $\overline{(6)}$.

But the last three rows of the table show that

(1) in the curve II (the rows of the table corresponding to $r = \dfrac{1}{2} + \sin 3\ \varphi$) the first positive loop (1) is situated in sector (—10°, 70°) the greatest $r = 1·5$), the next loop, which is negative $\overline{(2)}$ lies in sector (250°, 290°) (the greatest value of r is 0·5) etc. (see Fig. 77a).

(2) in the curve III there are positive loops only in sectors (—30°, 90°), (90°, 210°) and (210°, 330°) (Fig. 77b).

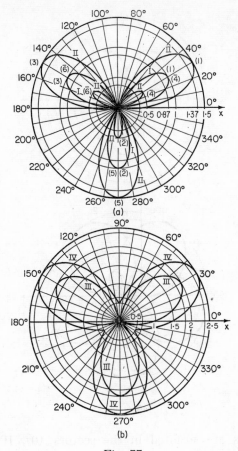

Fig. 77.

(3) in the curve IV the smallest value of r is 0·5 and the loops are of an "unfinished" form (Fig. 77*b*). The situation is entirely analogous as far as the curves

$$r = a + \sin\frac{5\varphi}{3} \text{ when } a = 0, \quad \frac{1}{2}, \quad 1, \quad \frac{3}{2}.$$

are concerned. Here it is convenient to change the value of the angle "φ" by 18° at a time (from 0° to 1080°). When $a = 0$, the first (positive) and the second (nega-

169

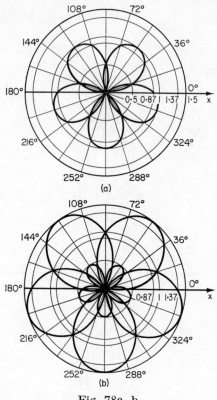

Fig. 78a, b.

tive) lobes are situated in the sectors (0°, 108°) and (288°, 396°) (Fig. 78*a*) and when $a = \frac{1}{2}$ the loops are situated in the sectors (−18°, 126°) and (306°, 378°) (Fig. 78*b*). When $a = 1$ there are five positive loops only in the sectors (−54°, 162°), (162°, 378°), etc.: the same, but with unfinished loops happens when $a = \frac{3}{2}$ (Fig. 78*c*).

In general, for the curve $r = \sin\left(\frac{(m\varphi)}{n}\right)$ the first positive loop is situated within the sector $\left(0°, \frac{180°n}{m}\right)$,

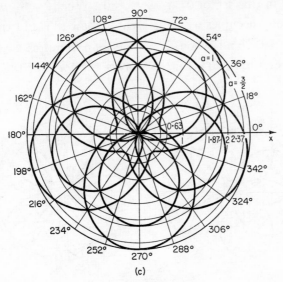

(c)

Fig. 78c.

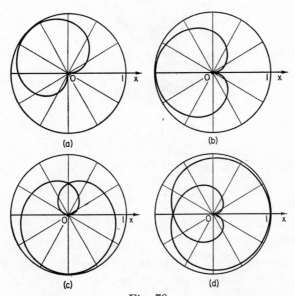

(a)

(b)

(c)

(d)

Fig. 79.

since in this sector $0° \leq \frac{m\varphi}{n} \leq 180°$. When $\frac{1}{2} < \frac{m}{n}$ $m < 1$, the loop occupies a sector greater than 180° but smaller than 360°, and when $\frac{m}{n} < \frac{1}{2}$ one loop requires a sector greater than 360° (Fig. 79 shows the form of the lobes for $\frac{m}{n} = \frac{2}{3}, \frac{1}{2}, \frac{1}{3}, \frac{1}{4}$)

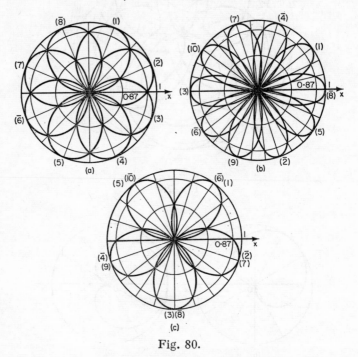

Fig. 80.

Regarding m and n as relatively prime, we consider the equation $r = \sin\left(\frac{m\Phi}{n}\right)$ The following cases may occur:

(1) m is even, n is odd. On varying φ from 0 to $360°\,n$, we obtain a closed rosette of $2m$ loops $\left(360°n \div \frac{180n}{m} = 2m\right)$ the last of which is negative; therefore, on varying φ further, we again traverse the rosette (in Fig. 80a, $m = 4$, $n = 3$).

(2) m odd, n even. On varying φ from 0 to $n \times 180°$ (a whole number of circles we obtain m loops $\left(n \times 180° \div \frac{180°n}{m} = m\right)$ but the last one is positive, therefore, on varying φ further from $n \times 180°$ to $n \times 360°$, we obtain loops, which are diametrically opposed to the ones already constructed; in total we again obtain a rosette of $2m$ loops (in Fig. 80b, $m = 5$, $n = 2$).

(3) m and n are odd. On varying φ from $0°$ to $n \times 180°$ m lobes are obtained, and since the last one is positive, therefore the next $(m + 1)$th loop is negative and coincides with the very first positive loop; on the whole, when φ varies from $n \times 180°$ to $n \times 360°$ we obtain all m loops which have been constructed already, but those which were positive in the first circuit become negative, and vice versa (in Fig. 80c, $m = 5$ and $n = 3$).

Lissajous Curves

Many interesting curves can be constructed in the Cartesian coordinates also. Particularly simple is the construction of curves, whose equations are given in the parametric form:

$$x = \varphi(t),$$
$$y = \varphi(t),$$

where t is a parameter.

In this case, the construction of the curve is preceded by the calculation of the values of x and y corresponding to values of the parameter t (increasing or decreasing) of sufficient degree of approximation. The points (x, y) are marked on the paper and joined by a smooth continuous line in order of t increasing. For example, for the equations $\begin{cases} x = \sin 2t \\ y = \sin 3t \end{cases}$ we have

t	$0°$	$15°$	$30°$	$45°$	$60°$	$75°$	$90°$	$105°$	...
x	0	0·5	0·87	1	0·87	0·5	0	—0·5	...
y	0	0·7	1	0·7	0	—0·7	—1	—0·7	...

Having continued this table to $t = 360°$ and marked in the points A_0, A_1, A_2,. ., A_{24} in the drawing, in accordance with the values of x and y obtained, it is easy to obtain the curve depicted in Fig. 81.

This is one of the so-called Lissajous curves, characterized by the general equations

$$\left. \begin{array}{l} x = a \sin mt, \\ y = b \sin n(t+\alpha). \end{array} \right\} \qquad (1)$$

If we take time as the parameter t, the Lissajous figures represent the result of compounding two simple harmonic vibrations occurring in mutually perpendicular directions.

Verify that the equations (I) $x = \sin 3t$; $y = \sin 5t$; (II) $x = \sin 3t$; $y = \cos 5t$; (III) $x = \sin 3t$; $y = \sin 4t$; (IV) $x = \sin (t - 45°)$, $y = \sin t$ are those whose curves are shown in Fig. 82. In addition to Lissajous curves,

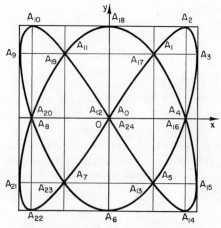

Fig. 81.

the drawing contains squares of side 2, inside which and touching whose sides, these curves are situated. In general, the curve (1) is situated inside a rectangle of sides $2a$ and $2b$.

In constructing Lissajous curves, it is useful to predetermine the points of contact of the curve with the sides of the corresponding rectangle, and also to draw beforehand auxiliary straight lines inside the rectangle:

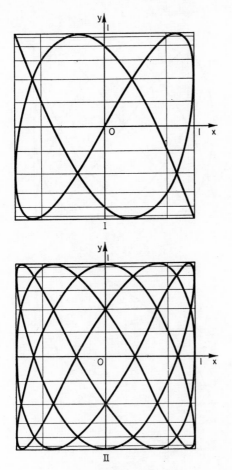

Fig. 82, I and II.

horizontal ones at a distance of $b \sin 15°$, $b \sin 30°$, $b \sin 45°$, $b \sin 60°$, $b \sin 75°$ from the origin of the coordinates, and vertical ones at distances $a \sin 15°$, $a \sin 30°$, etc., from the origin (regarding φ as varying $15°$ at a time and $\alpha = 0$).

A Lissajous curve is open if, for some value of the parameter t, the curve becomes "wedged" at a vertex

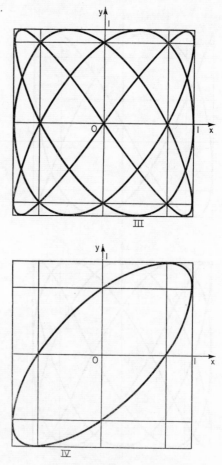

Fig 82, III and IV.

of the rectangle: as ϕ increases further, the point tra-
verses the same curve in the opposite direction (see,
e. g. the curve I in Fig. 82).

Find under what conditions the curve $(I)\ \alpha = 0$,
is open([68]). It is useful to investigate how the substitution of, say, equations $x = \sin 3t$; $y = \sin 5\,(t + 3°)$
for eqns. (I) transforms an open curve into a closed
one.

Cycloids, Epicycloids, Hypocycloids

Readers will derive great pleasure from a close
acquaintance with the cycloid, the epicycloid and the
hypocycloid — these are the names of the curves
described by the point M situated on the circumference
of a circle, radius r, rolling along a straight line (*cycloid*),
along the circumference of a stationary circle of radius
R, touching it on the outside (*epicycloid*) and along
the circumference of a stationary circle of radius R,
touching it on the inside (*hypocycloid*).

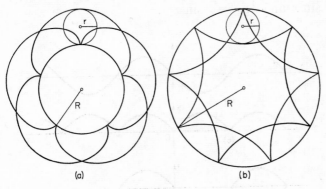

(a) (b)

Fig. 83.

The equations of these curves can be represented
in the parametric form, thus:

$$\text{cycloid} \quad \begin{cases} x = r\,(t - \sin t), \\ y = r\,(1 - \cos t), \end{cases}$$

177

epicycloid $\begin{cases} x = R[(1+m)\cos mt - m\cos(1+m)t], \\ y = R[(1+m)\sin mt - m\sin(1+m)t], \end{cases}$

hypocycloid $\begin{cases} x = R[(1-m)\cos mt + m\cos(1-m)t], \\ y = R[(m-1)\sin mt + m\sin(1-m)t], \end{cases}$

where $m = \dfrac{r}{R}$ (see [27], vol. I).

The book by G. N. Berman [4] is devoted to interesting properties of these curves, and it shows also the external form of epicycloids and hypocycloids for various values of m. In Fig. 83 an epicycloid (a) for $m = \dfrac{2}{5}$ and a hypocycloid (b) for $m = \dfrac{2}{9}$ are shown.

Interesting Broke n Lines

Great possibilities in varying the form of curves are opened up by the use of the sign of absolute value in equations of curves.

For example, the curves (a), (b), (c) and (d) in Fig. 84 correspond to equations $y = \sin x$, $y = |\sin x|$, $|y| = \sin x$ and $|y| = |\sin x|$.

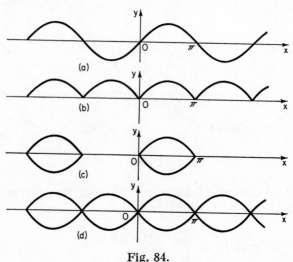

Fig. 84.

Construction of Curves

The curve corresponding to the equation
$$|x| + |y| = 1, \qquad (2)$$
represents a square $ABCD$ (Fig. 85a), since, for example, for the second quadrant, where $x < 0$ and $y > 0$, the eqn. (2) can be rewritten in the form $-x + y = 1$ and this is the equation of the straight line l, only a segment of which, lying in the second quadrant, is taken.

Show that to the equations

I. $|2y - 1| + |2y + 1| + \dfrac{4}{\sqrt{3}}|x| = 4.$

II. $|x| + |y| + \dfrac{1}{\sqrt{2}}\{|x - y| + |x + y|\} = \sqrt{2} + 1,$

III. $|x| + |y||-3|-3| = 1$

there correspond a regular hexagon, a regular octagon, and a "figure eight", as shown in Fig. 85b, c, d (see [35] pp. 194–195). [69]

Perhaps the reader would succeed in finding equations giving a regular dodecagon, sixteen-sided polygon, etc. The solution of this problem for regular pentagon, septagon, etc. is more complicated.

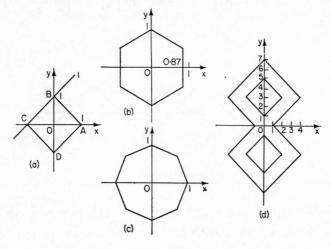

Fig. 85.

179

Lines, corresponding to the equations of form

$$y = m \arcsin [\sin k (x - \alpha)].$$

are also quite unusual.

It follows from the equation $y = \arcsin (\sin x)$ that

$$(1) -\frac{\pi}{2} \leqslant y \leqslant \frac{\pi}{2} \text{ and } (2) \sin y = \sin x,$$

When $-\frac{\pi}{2} \leq x \leq \frac{\pi}{2}$, these two conditions are satisfied by the function $y = x$. In the interval $\left(-\frac{\pi}{2}, \frac{\pi}{2}\right)$ its graph is the segment AB of the broken line shown in Fig. 86.

In the interval $\frac{\pi}{2} \leq x \leq \frac{3\pi}{2}$ we have $y = \pi - x$, since $\sin (\pi - x) = \sin x$, and in this interval $-\frac{\pi}{2} \leq \pi - x \leq \frac{\pi}{2}$. Here the graph is the segment BC.

Since $\sin x$ is a periodic function of period 2π, the broken line ABC, constructed in the interval $\left(-\frac{\pi}{2}, \frac{3\pi}{2}\right)$, is repeated once again in the regions $\left(\frac{3\pi}{2}, \frac{7\pi}{2}\right)$, $\left(\frac{7\pi}{2}, \frac{11\pi}{2}\right)$ etc.

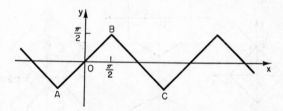

Fig. 86.

To the equation $y = \arcsin (\sin kx)$ there corresponds the broken line I with the period $\frac{2\pi}{k}$ (the period of the function $\sin kx$) (in Fig. 87, $k = 2$).

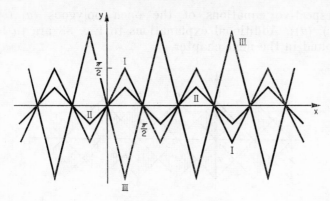

Fig. 87.

Adding to the right-hand side a factor m, we obtain the equation $y = m$ arcsin (sinkx), to which there corresponds, for $m > 0$ a broken line of the form *II* $\left(m = \frac{1}{2}\right)$, and for $m < 0$ a broken line of form *III* $(m = -2)$.

Finally, it is easy to see ([70]) that by displacing the latter lines to the right by a distance α, we obtain graphs of functions of the form

$$y - m \arcsin [\sin k(x-\alpha)].$$

We suggest that the reader verifies that the set of lines shown in Fig. 88 corresponds to the equation

$$\left[y - \frac{2}{\pi} \arcsin\left(\sin\frac{\pi x}{\pi}\right)\right]\left[y + \frac{2}{\pi} \arcsin\left(\sin\frac{\pi x}{\pi}\right)\right] \times$$

$$\times \left[y - \frac{2}{\pi}\arcsin\left(\sin\frac{\pi(x-1)}{2}\right)\right]\left[y + \right.$$

$$\left. + \frac{2}{\pi}\arcsin\left(\sin\frac{\pi(x-1)}{2}\right)\right] = 0$$

(the product equals zero only when one of the factors becomes zero. By making the first, second, third and fourth factors in turn become zero, we obtain the

181

respective equations of the open polygons (*a*), (*b*), (*c*), (*d*)]. Additional explanations to Fig. 88 are to be found in the next chapter.

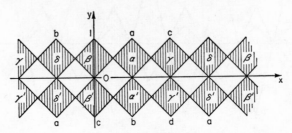

Fig. 88.

Ornate curves

Certain curves, whose equations contain very high powers of numbers, possess certain very interesting properties (see M. L. Frank's article in [38], 6th issue, 45–54).

For example, for the equation $y^{1000000} = x$ we have the table:

x	0	10^{-6}	1	10^{10}	$10^{100\,000}$	
y	0	± 0.999986	± 1	± 1.000023	± 1.2583	check!

from which it can be seen, that a considerable part of the curve does not differ to any practical extent from the broken line *LMNP* (Fig. 89).

Figure 90 shows the graph of the function $y = \sin x +\\ + 2 (\sin x)^{1000000}$

x	89°45	89°50	90°	
$2 (\sin x)^{1\,000\,000}$	0.00016	0.032	2	check!

It can be seen from the table, that the second term of the given function differs very little from zero, even when the value of x is quite close to 90° (also to 270°, etc), therefore there appear in the vicinity of certain points of the ordinary sine curve (*A*, *B*, etc.) very narrow upward peaks, which look like segments *AC*, *BD*, etc., when the curve is drawn to a comparatively small scale.

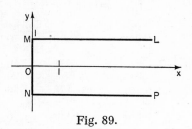

Fig. 89.

To the equation $y^{1000001} = \sin\frac{\pi x}{2}$ there corresponds a curve, differing very little from the broken curve *ABCDEFGHI* (Fig. 91) (as soon as $x = 0 \cdot 000001$, $y = 0 \cdot 999984$).

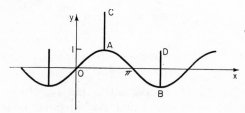

Fig. 90.

Terms containing smaller powers of some periodic function can be used to alter the form of other comparatively simple curves.

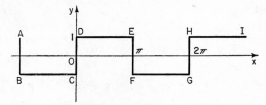

Fig. 91.

By using a table of values of sin x and the table below, which can be continued easily,

x	$\sin 3x$	$\sin^5 3x$	$\sin^9 3x$
0	0	0	0
5°	0·259	0·0001	0·000
10°	0·5	0·031	0·002
15°	0·707	0·178	0·044
20°	0·866	0·487	0·28
25°	0·966	0·84	0·73
30°	1	1	1

try to draw large scale graphs of the following functions

$$y = \sin x + \frac{1}{3} (\sin 3x)^5$$

$$y = \sin x - \frac{1}{2} (\sin 3x)^9.$$

The influence of additional terms on the form of a curve can be usefully studied also in polar coordinates.

The interesting article by N. F. Chetverukhin ([37], 1930, No. 5) contains some of the equations found by the German mathematician and naturalist Habenicht for the geometric forms found in the plant world.

For example, the curves, shown in Fig. 92 correspond to equations $r = 4 (1 + \cos 3\varphi)$ and $r = 4 (1 + \cos 3\varphi) + 4 \sin^2 3\varphi$.

The equations

$$r = 5 + 2 \cos \varphi + 3 \cos^{71} \varphi$$

and

$$r = 5 + 2 \cos \varphi + 3 \cos^{71} \varphi - \sin^2 18 \varphi \cos^4 \frac{\varphi}{2}$$

give the outlines of "a lilac leaf" and "a nettle leaf".

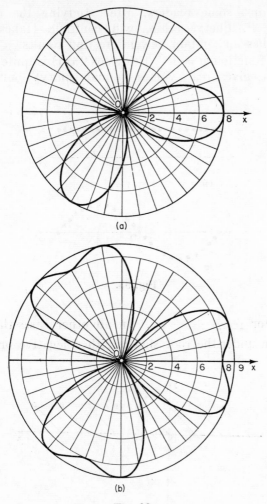

(a)

(b)

Fig. 92.

We advise the reader to construct the curve $r = 5 + 2 \cos \varphi$ first, and then to observe the influence of the term $3 \cos^{71} \varphi$ followed by the influence of the term $- \sin^2 18 \varphi \cos^4 \frac{\varphi}{2}$.

185

Perhaps, some readers, after studying the article by N. F. Chetverukhin, will continue Habenicht's researches and obtain new interesting results.

The function $y = 1 + \sqrt{(\log \cos 2\pi n x)}$ is quite peculiar. It gives real values (equal unity) only for

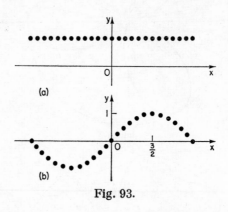

Fig. 93.

$x = 0$, or $\pm \dfrac{1}{n} \pm \dfrac{2}{n}$, ... In all other cases the expression under the root sign is either negative or imaginary.

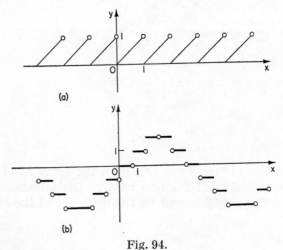

Fig. 94.

The graph of this function is a set of points situated above the axis Ox at a distance of 1 and at the distance from each other of $\frac{1}{n}$ (see Fig. 93a).

If the graph of the function $y = f(x)$ is known, then, by preserving only those points of this curve, whose abscissae are $0, \pm \frac{1}{n}, \pm \frac{2}{n}$, etc., we obtain the graph of the function $y = f(x) \left\{ 1 + \sqrt{(\lg \cos 2 \pi nx)} \right\}$ (see Fig. 93b, where $f(x) = \sin \frac{\pi x}{3}$ and $n = 4$).

The use of the function $f(x) = [x]$ (see § 2) helps to convert uninteresting functions into functions with graphs of quite original form. For example, Fig. 94 shows (a) the graph of the function $y = x - [x]$ and (b) the graph of the function $y = \left[\frac{6}{\pi} \text{ arc sin } \frac{(\sin \pi x)}{6} \right]$; the little hollow circles indicate that the corresponding points do not appear in the graph.

§ 29. Mathematical Borders

We apply the name mathematical border to a drawing characterized by some equation or inequality (or even a set of equations or a set of inequalities) in which a certain pattern is repeated many times.

For example, the set of inequalities

$$y > \sin x, \\ y < -\sin x \quad \Big\} \tag{1}$$

is satisfied by the coordinates of points, which lie above the sine curve (for them $y > \sin x$) and below the curve $y = \sin x$, at the same time, i.e. the region of solutions of the set (1) consists of the regions, shaded in Fig. 95.

Let us denote the first and the second factor in the inequality

$$(y - \sin x)\,(y + \sin x) < 0. \tag{2}$$

by $f_1(x, y)$ and $f_2(x, y)$.

This inequality is satisfied by the coordinates of points, for which $f_1(x, y) > 0$ and $f_2(x, y) < 0$ at the same time (they lie in the shaded region in Fig. 95) and by the coordinates of points for which $f_1(x, y) < 0$ and $f_2(x, y) > 0$ (they fill the regions marked by asterisks.)

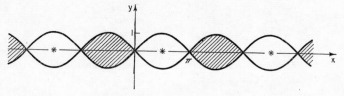

Fig. 95.

188

The whole region of solutions of the inequality (2) consists of the shaded areas and the starred areas. If in the inequality

$$\left[y - \frac{2}{\pi} \arcsin\left(\sin\frac{\pi x}{2}\right)\right]\left[y + \frac{2}{\pi} \arcsin\left(\sin\frac{\pi x}{2}\right)\right] \times$$

$$\times \left[y - \frac{2}{\pi} \arcsin\left(\sin\frac{\pi(x-1)}{2}\right)\right] \times$$

$$\times \left[y + \frac{2}{\pi} \arcsin\left(\sin\left(\frac{\pi(x-1)}{x}\right)\right)\right] < 0 \quad (3)$$

we denote the factors in the left-hand side by φ_1, φ_2, φ_3, φ_4 respectively, then the inequality (3) will be satisfied by the coordinates of points at which either one of the factors is negative and the rest positive (these are, for instance, regions α in Fig. 88, where $\varphi_1 < 0$ and φ_1, φ_2, $\varphi_3 > 0$, regions β, where $\varphi_4 < 0$ and φ_1, φ_2, $\varphi_3 > 0$, etc.) or three factors in the left-hand side are negative and the fourth one is positive (regions α' where φ_1, φ_2, φ_4, < 0 and $\varphi_2 > 0$, regions β', where φ_1, φ_2, $\varphi_4 < 0$ and $\varphi_3 > 0$, etc.).

The whole region of solutions of inequality (3) consists of squares, shaded-in in Fig. 88.

By increasing the number of factors in the left-hand side of the inequality, selecting them in such a way, that the curves and broken lines at which they become zero are shifted relative to each other by this or that distance, by taking a set of simple inequalities in place of one complex inequality, etc., it is possible to vary the form of drawings obtained very considerably.

We leave it to the reader to verify that the regions of solutions of

(1) the inequality

$$\{y^2 - \arcsin^2(\sin x)\}\left\{y^2 - \arcsin^2\left[\sin\left(x + \frac{\pi}{6}\right)\right]\right\} < 0 : \quad (4)$$

(2) the inequality

$$\Phi_{-2}\, \Phi_{-1}\, \Phi_0\, \Phi_1\, \Phi_2 < 0,$$

where $\varPhi_k = \varPhi_k(x, y) = y^2 - \arcsin^2\left[\sin\left(x + \dfrac{k\pi}{8}\right)\right]$ and $k =$

$= -2, -1, 0, 1, 2;$

(3) the inequality

$$[y^2 - \sin^2 x]\left[y^2 - \sin^2\left(x + \dfrac{\pi}{6}\right)\right]\left[y^2 - \sin^2\left(x - \dfrac{\pi}{6}\right)\right] < 0; \quad (6)$$

(4) the set of inequalities

$$\begin{cases} y^2 - \left(\dfrac{2}{\pi}\right)^2 \arcsin^2\left(\sin\dfrac{\pi x}{4}\right) < 0, & (7') \\[4mm] y^2 - \left(\dfrac{4}{\pi}\right)^2 \arcsin^2\left[\sin\dfrac{\pi}{2}\,(x-1)\right] < 0: & (7'') \end{cases}$$

and (5) the set of inequalities

$$\begin{cases} y^2 - \left(\dfrac{16}{\pi}\right)^2 \arcsin^2\left(\sin\dfrac{\pi x}{8}\right) < 0, & (8') \\[4mm] y^2 - \left(\dfrac{16}{\pi}\right)^2 \arcsin^2\left[\sin\dfrac{\pi(x-1)}{8}\right] < 0, & (8'') \\[4mm] y^2 - \left(\dfrac{16}{\pi}\right)^2 \arcsin^2\left[\sin\dfrac{\pi(x-3)}{8}\right] < 0, & (8''') \\[4mm] y^2 - \left(\dfrac{16}{\pi}\right)^2 \arcsin^2\left[\sin\dfrac{\pi(x-6)}{8}\right] < 0 & (8'''') \end{cases}$$

consist repectively of the shaded figures (*a, b, c, d, e*). In Fig. 96 (in Fig. 96*d*) a dotted line marks the wide rhombi, which give the solution of the inequality (7′) only and the tall rhombi, which give the solution of the inequality (7″) only. In Fig. 96*e*, the large rhombi I delimit the region of solutions of inequality (8′), the rhombi II delimit the region of solutions of the inequality (8″) etc.; the common part of these regions gives a chain of little rhombi.

By making use of inequalities with high powers of variables, it is possible to increase the variety of mathematical borders. For example, to the inequality

$$\left(y^{1000001} - \sin x\right)\left[(2y)^{1000001} - \sin\left(x - \dfrac{\pi}{2}\right)\right] < 0 \quad (9)$$

Mathematical Borders

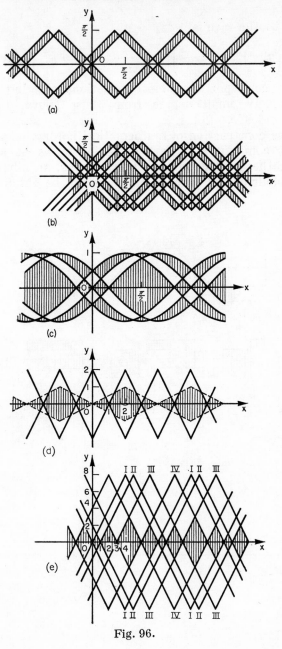

Fig. 96.

and to the inequality

$$\left(y^{1000001}-\sin\frac{\pi x}{4}\right)\left(y-\frac{4}{\pi}\arcsin\sin\frac{\pi x}{4}\right)<0 \quad (10)$$

there correspond regions of solutions differing to an extremely small degree from those shown in Fig. 97*a* and 97*b*.

The construction of mathematical borders may serve as a competition theme in senior forms at school; by gathering together the best borders, an album full of colourful and fanciful drawings can be obtained.

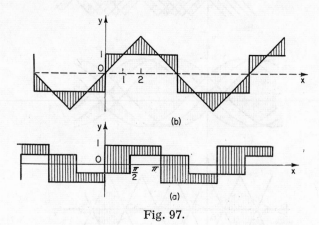

Fig. 97.

§ 30. Models of Polyhedra

Suppose the opposite edges of a regular tetrahedron, the opposite vertices of a cube and of a regular icosahedron (where $AB \perp \varrho$ and $CD \perp \varrho$), the opposite faces of a regular octahedron and a regular dodecahedron (Fig. 98) lie in two parallel planes, ϱ and σ.

If we intersect all these polyhedra by a plane parallel to ϱ and σ and half-way between them, we obtain as cross-sections, a square, two regular hexagons and two regular decagons.

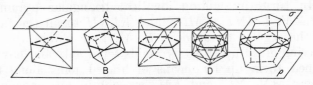

Fig. 98.

If we prepare models of the two halves of each polyhedron and join them by a hinged join (by sticking them together along a common edge with a narrow strip of strong paper or cloth) we obtain visual models of the cross-sections and the parts into which the polyhedra are split by the "medial plane".

Figure 99 gives plane developments of the dodecahedron and icosahedron.

Models, showing the transition from one regular polyhedron to another are also interesting and instructive. It can be seen from Fig. 100a that a regular octa-

193

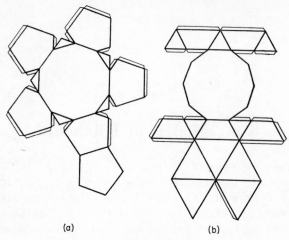

(a)   (b)

Fig. 99.

hedron *ABCDEF* can be completed to form a regular tetrahedron *KLMN* by adding on to each of its four faces a little tetrahedron.

In turn, by adding identical triangular pyramids *KLMV*, *LMNT*, *KMNS* and *KLNU* to all the faces of the regular tetrahedron *KLMN*, we obtain a cube. Having glued up these pyramids separately and having joined them up by means of hinged joins along the edges *MN*, *NL*, and *LM*, we obtain an "envelope" which can be tucked round the tetrahedron *KLMN* to form a cube (Fig. 100*b*).

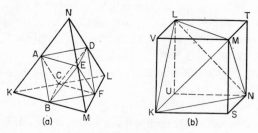

(a)   (b)

Fig. 100.

194

Models of Polyhedra

For the cube, it is possible to glue together an envelope of identical lids *ABCDKL, ADEFMN*, etc., transforming it into a regular dodecahedron (Fig. 101*a*). Figure 101*b* shows separately the unfolded lid of this kind; the obtuse angles of the triangles and trapeziums are 108° each and $b = \dfrac{a}{2 \cos 36°} \simeq 0 \cdot 618a$.

Finally, out of the twelve polyhedra, shown on Fig. 101*c*, it is possible to glue together an envelope, which transforms a regular dodecahedron of edge *a* into a regular icosahedron circumscribed about the dodecahedron. Each little lid is bounded by a regular pentagon *ABCDE* of side *a*, 5 isosceles triangles (*AKE, EMD,* etc.) of lateral side $b \simeq 0.535a$ and 5 identical quadrilaterals (*EKSM, DMSL,* etc.) $EK = EM = b; KS = = SM \simeq 0.927a; KEM = 120°, EKS = EMS = 90°$.

Using models, it is possible to become acquainted with *prismatoids* of various forms — this is the name given to convex polyhedra, whose bases are polygons of any shape, which lie between parallel planes, and whose side faces are triangles (or, in some cases, trapeziums)

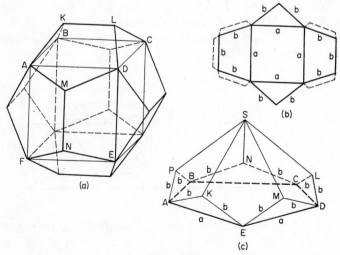

Fig. 101.

all of whose vertices coincide with the corresponding vertices of the bases. Figure 102 shows a prismatoid *ABCDKLM* with triangular and quadrilateral bases, a prismatoid *EFHQR* with triangular and "bi-angular" (segment *QR*) bases, and a so-called *regular antiprism* — this is the name of prismatoids, whose bases are identical regular *n*-sided polygons, whose side-faces are equilateral triangles, and in which the "line of centres of bases" $O_1 O_2$ is perpendicular to ϱ and the bases are rotated relative to each other by an angle $\dfrac{\pi}{n}$.

It is easy to construct a plane development of a prismatoid if we are given its projection on to the plane ϱ (or σ) and its height h (Fig. 103).

The bases of the prismatoid appear undistorted in the projection; in order to determine, for example, the true shape of the face *CDM* it is necessary to draw $M'P \perp CD$ and measure off $PM_1 = \sqrt{(M'P^2 + h^2)}$ (see Fig. 102). The true shape of other side faces adjacent to the base *ABCD* is determined similarly, and so are the faces adjacent to the base *KLM* (they are marked in dotted lines in Fig. 103).

We cut out figures $AL_1 BL_2 CM_1 DK_1 A$ and $K' A_1 L'C_1 M'D_1 K'$ (having left enough material at the edges for glueing) and then we easily glue together the prismatoid itself.

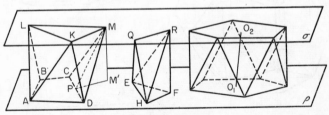

Fig. 102.

Construct plane developments and glue together models of several irregular prismatoids, and also of several regular antiprisms (for $n = 5, 6, 7, \ldots$).

As can be seen from Fig. 104, a cube may be constructed out of three identical pyramids; *LABCD*, *LMNCD*, *LAKND*, which have a common edge *LD*

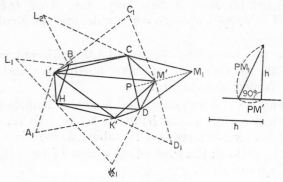

Fig. 103.

(the diagonal of the cube) and square bases *ABCD*, *NMCD*, *AKND*; hence it follows — without using the formula for the volume of a pyramid — that the volume of a pyramid with a square base, whose side edge equals the side of the base *a* and is perpendicular to the plane of the base, equals $\frac{a^3}{3}$.

By rotating the cube about the diagonal *LD* through 120° and through 240°, we see that the pyramids *LABCD*, *LNMCD*, *LAKND* are transformed into each other.

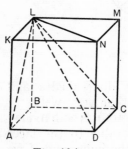

Fig. 104.

Make a composite model of a cube out of models of these pyramids, by joining the first one with the second one by means of a hinged join along the edge *LC*, and the first one with the third one, in the same way, along the edge *LA*.

If we fill space with identical cubes in such a way that any two neighbouring cubes have a common face, and if we mentally colour them white and black in chesslike order, then on passing planes through the edges of each black cube *ABCDKLMN* (Fig. 105) so that they pass at the same time through the centres O_1, O_2 of the neighbouring cubes, (say, the plane O_1KN), we shall have split up each white cube into six identical pyramids.

If we join on to each black cube the adjacent white pyramids, we find that we have filled space with so-called *rhombic dodecahedra*.

A rhombic dodecahedron has 8 angles bounded by three faces and 6 angles bounded by four faces, and each of its twelve faces is a rhombus, whose diagonals are in the ratio $1 : \sqrt{2}$. Prove that planes of form O_1KN and KO_2N coincide.

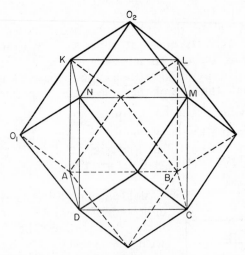

Fig. 105.

Construct 5–6 identical rhombic dodecahedra and it will be seen that they fit each other very well.

If the cube *ABCDKLMN* (Fig. 106) is subdivided into eight little cubes by planes parallel to the faces and passing through the centre of the cube, and if we cut off from each little cube that half of it which is adjacent to the corresponding vertex of the large cube (for instance, the septahedron *NXYZPQRSTU* adjacent to the vertex *N*), we are left with a polyhedron of 6 square and 8 regular hexagonal faces, composed as if of 8 little half-cubes with a common vertex in the centre *O* of the large cube.

An exactly similar polyhedron can be constructed out of 8 half-cubes grouped about the point *N*, if it is supposed that the operation described above is performed on each of 8 large cubes with a common vertex *N*.

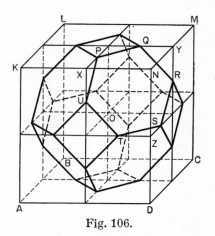

Fig. 106.

Therefore, it is possible to fill space, without any gaps, with the fourteen-faced polyhedra shown in Fig. 106. This can be verified visually by making up 5 of such polyhedra of identical dimensions.

It is possible to make folding models of certain polyhedra, using firm cardboard. For instance, cut 4 regular triangles out of cardboard and hinge them together in two pairs by glueing firm material along the straight line *AB* and along the straight line *CD* (Fig. 107*a*).

If we join the rhombi obtained by means of a thin taut elastic band *ONMLKVUTSRQPO*, threaded through openings *M* and *L*, *V* and *U*, *S* and *R*, *P* and *Q*, then the elastic band, by becoming shorter, tends to change the plane figure (*a*) into the tetrahedron (*b*) (points *U* and *V* lie on the median of the triangle *ABC*, etc.).

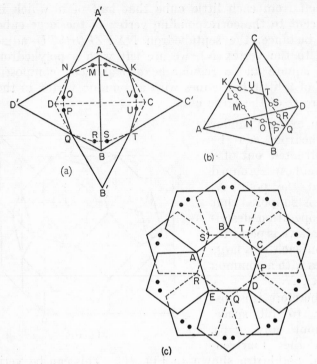

Fig. 107.

In order to obtain folding models of a regular dodeca-hedron we superimpose a rosette of six regular card-board pentagons, five of which are joined by a hinge-like join with the sixth central pentagon *ABCDE* (Fig. 107c), over a similar rosette with central pentagon *PQRST* and an elastic band is threaded tautly through the twenty holes shown in the drawings.

PROBLEMS: 1. By passing planes through each of the twelve edges of a cube so that the planes make equal angles with the faces bounding the corresponding angle, we arrive at a rhombic dodecahedron (Fig. 105).

By passing planes through the edges of a rhombic dodecahedron, so that they are equally inclined to its corresponding faces, we arrive at the polyhedron S_1

circumscribed about the rhombic dodecahedron; subsequently, it is possible to obtain S_2, from S_1 in the same way, and from $S_2 - S_3$, etc.

Attempt to determine the form of the faces of S_1 and S_2 and to construct their models.

Similar problems can be set, using in place of a cube as base some other polyhedron (a regular dodecahedron, a regular pentagonal antiprism etc.).

What is the result, it we take a regular tetrahedron or a regular octahedron as the starting polyhedron?[n].

2. If we fill space with identical parallelepipeds of

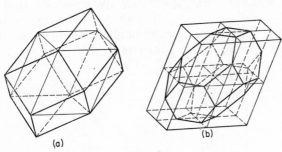

(a)   (b)

Fig. 108.

arbitrary shape, coloured as for chess, then, on joining on to each black parallelepiped 6 pyramids whose vertices are at the centres of the neighbouring parallelepipeds, we obtain dodecahedra, filling space without gaps (Fig. 108a). On cutting off 8 halves of little parallelepipeds from each large parallelepiped, we obtain 14-faced polyhedra, also capable of filling space (Fig. 108b).

Try to construct plane developments of the 12-faced and 14-faced polyhedra obtained as above, if you are given the height h of the parallelepiped and its projection on to its base plane.

Having glued together several polyhedra of each type, verify that your models do fit in with each other exactly.

§ 31. Pastimes with a Sheet and a Strip of Paper

A triangular sheet of paper enables us to demonstrate easily, without using geometric instruments, that the three bisectors of angles (the three medians, the three altitudes, the three perpendiculars at mid-points of sides) of a triangle are concurrent.

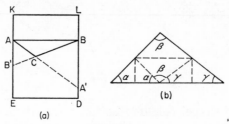

Fig. 109.

Indeed, any of the lines indicated can be easily constructed by folding the sheet in an appropriate manner. If the triangle ABC is obtuse, then, in order to obtain the centre of the circumscribed circle, we must take a sheet in the form $AKLBC$ (Fig. 109a); and in order to obtain the orthocentre we must take a sheet in the form $ABDE$, with continuations of sides AC and BC marked in by folds in the paper.

It is even simpler to prove the theorem about the sum of the angles of a triangle (Fig. 109b).

It is also possible to divide approximately — without drawing instruments — any angle ABC into three equal parts, for which it is sufficient (Fig. 110a) to

202

fold the paper along the line *BK*, which passes through the vertex of the angle, in such a way, that angles *KBA′* and *A′BC* are equal. With some practice, it is possible to do it freehand quite accurately.

Regular Polygons

If a strip *ABCD* with parallel edges (Fig. 110*b*) has $< ABC = 90°$, then, on folding it along the straight lines, *BK, KL, LM, MN,* . . ., it is possible to close the resulting "letter-cum-envelope" by means of the last "incomplete triangle" and thus obtain the shape of an equilateral triangle. If the strip is only half-folded along the lines *BK, KL,* . . ., then the side surfaces of various regular antiprisms may be constructed out of it (see § 30).

A strip of paper with parallel edges can be used to construct a regular pentagon, by tying the strip in a knot (Fig. 110*c*) and carefully pulling the knot tight, then pressing it flat in the form of a regular pentagon *KLMEF* (Fig. 110*d*).

If we then push the strip *EFCD*, bending it along the straight line *EF*, underneath the trapezium *KLMF*,

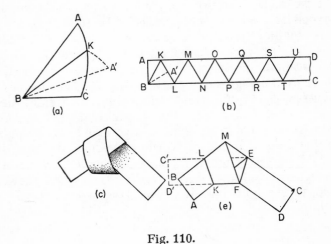

Fig. 110.

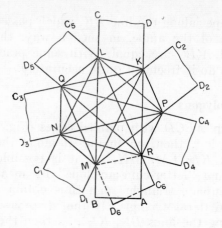

Fig. 111.

we obtain a "letter-cum-envelope" in the shape of a regular pentagon. When viewed against the light, and if made of fairly fine paper, this figure clearly shows a regular pentagonal star with a dark middle.

By folding a sufficiently long strip $ABCD$ along a straight line KL $\left(< AKL = \frac{4\pi}{7} \right)$, we obtain the first three vertices, K, L, M, of a regular septagon (in Fig. 111 we must mentally lengthen the strip considerably beyond the "curved boundaries" CD, then C_1D_1 etc.).

By folding the strip, consecutively, along the straight line MN (so that NC_2 passes through point K) along the straight line KP (so that PD_3 passes through the point N), along the straight line NQ (so that QC_4 passes through the point P) along the straight line PR (so that RD_5 passes through the point Q and PC_5, through the point L) and finally, along the straight line QL (as a result of which the edges of the strip should pass through the points M and R), we obtain a regular pentagon $MNQLKPR$, and when viewed against the light, it shows several septagonal stars.

Möbius' Strip

Bring into close proximity the two ends, *AB* and *CD* of a strip of paper, give it a twist (so that the point *C* coincides with *B* and *D* with *A*) and glue together the ends *AB* and *CD* while the strip is still twisted (Fig. 112). This gives the so-called Möbius strip — a one-sided surface, which cannot be coloured differently on each side; having started applying paint at some point and proceeding along the strip, we eventually arrive at the starting point on the opposite side of the paper.

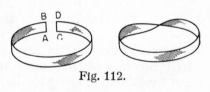

Fig. 112.

When a Möbius strip is cut along a line, equidistant from its edges, it does not fall apart, but is transformed into a "fully twisted ring", and the latter, on being cut along its median, gives two rings, twisted twice and interlinked in a fantastic manner. This rather striking property of a Möbius strip can be used as a basis for a trick to impress and mystify unsophisticated audiences.

Readers might well carry out a systematic investigation of the behaviour of a Möbius strip, and of repeatedly twisted paper rings, when they are cut along their median, or along two lines equidistant from each other and from the corresponding edges, etc.

Greater details of the properties of the Möbius strip are given in [6], pp. 117–132.

The Construction of a Regular Icosahedron

Take a strip of firm paper of width 10–12 centimetres, fold it in accordance with Fig. 110b and cut out several equilateral triangles. Each of them can be accurately subdivided freehand into sixteen equilateral triangles, folding the paper first along the lines *M'K'*, *K'L'*, *L'M'*, which join the mid-points of the sides,

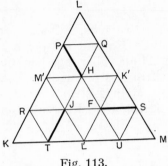

Fig. 113.

(*L* is made to coincide with the mid-point *L'* of side *KM'* and the paper is pressed along the line *M'K'*, etc.), then along the lines *PQ*, *RS*, *TQ*, etc. (see Fig. 113).

If we slit the paper along segments *PH*, *FS*, *TI* then, by rotating, say, the rhombus *PLQH* about the vertex *H* in an anti-clockwise direction, in such a way that the triangle *HQP* becomes situated underneath the triangle *HM'P*, and by folding it along its diagonal *PQ* until the vertex *L* coincides with point *H*, we obtain an angle bounded by five faces, having vertex *H* (kept well together by the bent triangle *PQL*, particularly if a drop of glue is placed at its centre beforehand).

Having constructed, in a similar way, angles of 5 faces with vertices at points *I* and *F*, we obtain a model of half the surface of a regular icosahedron.

Using 4–5 halves of icosahedra, sliding them on to each other in such a way that each two halves in direct contact have 4 faces in common, it is easy to obtain a sufficiently stable model of a regular icosahedron.

The reader might try constructing stable models of a regular tetrahedron, octahedron and icosahedron out of several strips of paper of the same width, subdivided into equilateral triangles (see Fig. 110*b*), and also a model of a regular dodecahedron out of several strips of paper each of which is "tied" in regular pentagons situated close to each other (see Fig. 110*d*).

§ 32. The Four-Colour Problem

Suppose we have to colour arbitrary regions of a plane or a spherical surface (for instance, the political map of the world) in such a way, that no neighbouring regions (i.e. having a common boundary, no matter how short) are coloured in the same way. Regions that touch at one or several points are not regarded as neighbouring and can be painted the same colour.

Experience shows that this can always be done by means of at most four colours, if, of course, we abandon the custom of painting all water areas on maps, which are also regarded as separate regions, the same colour.

It may happen, of course, that on painting the first few regions in the wrong order, a fifth colour may become necessary (for example, to paint over the central region in Fig. 114a where the numbers of colours used are marked in), but, by colouring the first regions in a different order, we can always succeed in using four colours only (Fig. 114b).

However, no one has succeeded in proving rigorously that it is impossible to subdivide a plane in such a way, that a fifth colour would be required, although it has been proved that 5 colours are always sufficient. ([24], pp. 90–101).

It is interesting that the surface of a torus (Fig. 114c) can be subdivided into as many as 7 regions each of which borders on all the others, and so, in general, we cannot colour a map on torus with less than 7 colours. However, it has been proved that it is impossible to subdivide the surface of a torus into regions in such a way that an eighth colour is required to colour all regions in all the various ways.

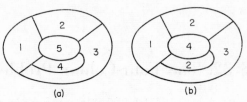

Fig. 114.

It is easy to see that in the case of space regions, completely different circumstances arise, namely, that any number of regions can be taken, each of which has a border with all the other regions in the form of a portion of some surface. It is sufficient to take two rows of long blocks laid on top of each other in different directions (Fig. 114c) and join up each two blocks of the same number into separate regions.

We shall indicate several problems related to the 4-colour problem.

1. Having glued together a model of a body (e) topologically equivalent to the torus, subdivide its surface into seven regions, each of which borders on six others.

2. A large "bridge" is situated in the middle of a courtyard. Subdivide the territory of the courtyard into seven sectors, using the upper surface of the bridge as well as the area under the bridge, so that any two sectors border on each other. Show the solution on a model (e) consisting of a sheet of paper, with a strip of paper stuck on to it in the form of a bridge.

3. Prove([72]) that an arbitrary number of straight lines subdivides a plane into regions, for the painting of which two colours are sufficient. What would the situation be in a space, which has been subdivided into regions by arbitrary planes?

4. Prove([73]) that there exist 8 tetrahedra (not necessarily regular ones) which can be distributed in space so that each of them has a common border with every other one, in the form of a portion of a face, without degeneration into a line or a point.

5. It can be seen from Fig. 114*g* that the territory of an island can be subdivided into six regions and they can be distributed among five states in such a way that any two states have bordering possessions. If it is required that the possessions of any state be coloured in one way, it is inevitable that five colours must be used.

Try to find, for $m = 6, 7, 8, 9, \ldots$, the smallest number n of regions into which an island has to be subdivided, so that when the regions are distributed among m states, any two states have bordering possessions.

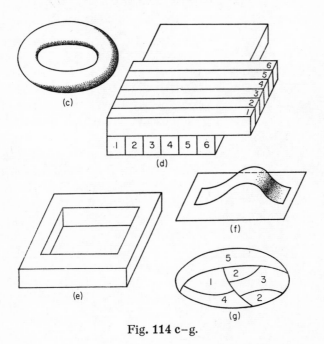

Fig. 114 c−g.

Perhaps you may find a formula of the form

$$n_{least} = f(m).$$

It is also possible, having subdivided the island into complicated and fancifully shaped regions, to consider the distribution of the regions amongst the greatest possible number of states, m, so that any two states have bordering possessions.

§ 33. Drawing Figures at one Stroke of the Pencil

It is possible to trace at one stroke, that is without lifting the pencil off the paper and without retracing any of the lines, a fairly complex figure (Fig. 115a), consisting of a number of "nodes" joined by "paths". (The paths may be curved (Fig. 115d) but it is impossible to trace out the very simple figure (b) in this way.)

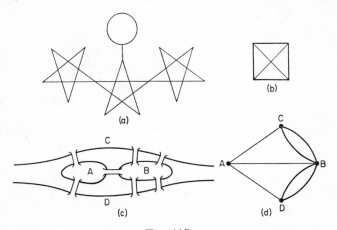

Fig. 115.

Let us call a point, from which there issue k paths, a node of the kth order, or a "k-node". Thus, the figure (a) has 2 nodes of the third order and 17 nodes of the fourth order. Evidently, any intermediate point on any of the paths can be considered as a node of the second order.

In tracing the figure in pencil we shall distinguish the *initial, final* and *intermediate nodes.*

THEOREM. *If a figure can be drawn at one stroke, none of its odd nodes can be an intermediate one.*

Indeed, a "$(2m + 1)$-node" can only be an initial one or a final one, since by passing through it m times we use $2m$ paths, and it is only possible to start or to complete the figure along the remaining path.

It follows from this theorem that it is impossible to draw a figure, possessing more than two odd nodes, at one stroke. In particular it is impossible to draw figure (*b*), which has four triple nodes.

Let us discuss the well known Euler problem about the seven bridges of Königsberg: is it possible to cross each of the seven bridges, connecting islands A and B with each other and with the river banks C and D once only? (Fig. 115*a*.)

Substituting points for the banks and the islands and lines for the bridges, we reduce Euler's problem to the question of drawing figure (*d*) at one stroke, which is impossible, as it has four odd nodes.

Prove([74]) that the number of nodes of odd order is even for any figure.

If we can pass from any point of a figure to any other point of it along paths belonging to it, the figure is called *connected.*

It can be proved that any connected figure without odd nodes, or with two odd nodes, can always be drawn at one stroke ([25], ch. 8).

The proof and the finding of a method of drawing of a figure which is drawable are based on the fact that when the tracing out of the figure is unsuccesful, and not all the lines of the figure have been included, it is always possible to find a point common to both the paths traced out and the paths left out in the tracing. It is possible to include a closed path along the untraced paths, beginning and ending at that point, into the first

variant of drawing, which diminishes the number of
paths untraced. This should be repeated until all the
paths are traced out at once.

Obviously, when two odd nodes are present one must
be a starting point and the other one a final point;
if all nodes are even, any point of the figure may be
taken as a point of departure, and the tracing will be
completed at that point.

If we agree that every line of the figure can be traced
out exactly twice (that is every path joining two
nodes is mentally exchanged for two paths) the order
of each node is doubled, and any connected figure can
be drawn at one stroke.

The double tracing of a figure can be completed
immediately if Tarry's rule is utilized ([25], p. 239–442):
after reaching a certain node along the path *l* for the
first time, this path *l* should be avoided for a return
visit into this node until this is possible again, that is
to pass along this path for the second time after all
the paths issuing from this node have been used twice;
at the same time, it is not possible to leave a node
twice along the same path (the second journey along
any path is made in the opposite direction.)

In Fig. 116, 16 numbered journeys are made, in
accordance with Tarry's rule; here the curved arrowheads show the paths of first-time approach to any
unvisited node (this path can
be used to move out of the node
only after all the approaching
paths have been used twice).

For any second approach to
a node, the corresponding path
is marked by a simple ar-
rowshead. Such a path can
be used for a second journey (in the opposite direction) at any time. For
example, in place of the
fifth move, the segment *M*

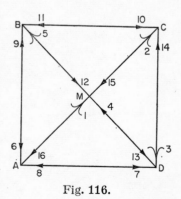

Fig. 116.

to D may be traced out, but no other move can be substituted for the eight one, since one journey has already been completed, along DM from the node D (the fourth move) and the curved arrowhead on DC shows that this path is taken from the node D last.

The Tarry's rule is of fundamental significance in the problem about mazes. Any maze can be regarded as a set of points (platforms, rooms, etc.) connected by lines (roads, corridors, etc.).

It follows from Tarry's rule that any connected maze (i. e. a maze without inaccessible points) can always be traversed by a person unacquainted with its plan in such a way that all paths are traversed twice. For that, in addition to "curved and simple arrowheads", one should mark paths with "halt sign" warning against retracing a passage for the second time in the same direction.

§ 34. Hamilton's Game

In the year 1857 the Irish mathematician Hamilton proposed a game, which he called "a journey around a dodecahedron", consisting of a journey from vertex to vertex of a regular dodecahedron, on condition that movement can take place only along the edges, and no vertex can be visited more than once.

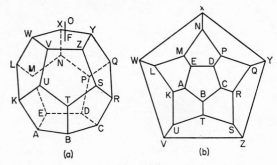

(a) (b)

Fig. 117.

If a central projection of the dodecahedron is made, i. e. its vertices and edges (Fig. 117a) are projected onto a plane passing through the face *ABCDE*, from the point *O*, which is situated at a small height above the centre of the face *VWXYZ*, the configuration (b) is obtained.

One of the problems in Hamilton's game is the construction of closed circuits connecting all twenty vertices. It seems (see [25] or [30]) that, having made any four moves from some vertex, it is possible to visit all remaining vertices and return to the original vertex with the twentieth move.

The question can be set about the number of circuits possible on being given, say, the first three vertices and the last one (depending on how they are placed, 0, 1, 2, 4 or 6 circuits are possible) or the first two and the last two vertices, etc. Here, we may be forbidden to visit certain vertices or to use certain edges.

A game may be considered involving two participants marking in the links of a "polygon of the circuit", one of the participants aiming at completing the circuit (not necessarily a closed one) of all the twenty vertices, while the other participant aims at creating a situation in which some of the vertices are left unvisited (for example, the polygon *SZYXNPDCRQ*).

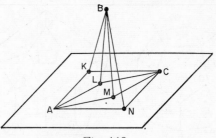

Fig. 118.

The polygon can be drawn by the players by placing numbered white and black pawns in turn in the corresponding vertices. The game becomes more complex and interesting if, instead of the configuration (*b*) we take a wooden dodecahedron with pins at the vertices on to which one can push well fitting rings.

Perhaps the reader may succeed in constructing a theory of the game, by proving the inevitability of defeat of one of the players if his opponent plays correctly (it is necessary, of course, to consider two alternatives, one when the player creating the impasse plays white, the other when he plays black).

Journeys around other polyhedra can be substituted for the journey around the regular dodecahedron. It is interesting for example, that it is impossible to visit all vertices of a rhombic dodecahedron while observing the rules of Hamilton's game (see § 30), since the rhombic dodecahedron has six angles bounded by four faces and eight angles bounded by three faces, and any edge joins different kinds of vertices.

Figure 118 shows a space configuration in which there are open circuits, but closed circuits are impossible (Why?) ([75]).

Try to find a polyhedron possessing the same property.

Hamilton also put forward another game — a journey round the faces of a polyhedron, in which the passage from one face to another is only permitted if the faces have a common edge.

We suggest that the reader verifies([76]) that the paths followed in Hamilton's second game as carried out on a regular octahedron and on a regular icosa-

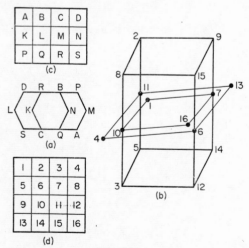

Fig. 119.

217

hedron are exactly the same as the paths followed in the first game as carried out on a cube and on a regular dodecahedron respectively.

The problem about a chess-knight in § 19 can be regarded as a problem of visiting every point of a given arrangement of points according to the rules of Hamilton's game; it is sufficient to replace squares of the board by points, perhaps situated differently, and to join up the points corresponding to knight's moves on the chess-board by lines.

Figure 119 shows a plane (*a*) arrangement and a space (*b*) arrangement corresponding to a "3,4-board" (*c*) and a "4²-board" (*d*).

§ 35. Arranging Points on a Plane and in Space

Problems of this type are usually "materialized" i. e. points are replaced by coins, nuts or other objects.

Let us quote a few typical problems:

1. Ten coins are arranged in the form of an equilateral triangle (see Fig. 120a). By moving three coins to new places, obtain a new equilateral triangle.

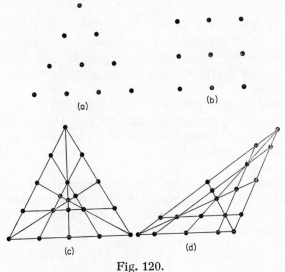

Fig. 120.

2. It can be seen from Fig. 120b that 9 coins can be distributed in 8 rows of 3 coins in each, i. e. it is possible to indicate 8 straight lines, each passing through the centres of 3 coins.

Move two coins[78] in such a way that the 9 coins become arranged in 10 rows each of 3 coins.

3. Figures 120c, d, show two fundamentally similar but outwardly different arrangements of 19 coins in 9 rows of 5 coins each.

Arrange 19 coins in 10 rows, 5 coins in each row[79].

Organize among your friends a competition in finding arrangements of n points in m rows, p points in each row, such that the ratio $\frac{mp}{n}$ is the greatest possible (in order to be definite, n should not be allowed to exceed a certain number n_0).

Attempt to investigate problems similar to the construction of magic squares. For example: can integers from 1 to 19 be placed at the nodal points of the configuration (c) of Fig. 120 in such a way, that the sums of numbers lying in any of the 9 straight lines are the same.

4. Arrange[80] 6 points in a plane, so that any 3 points are the vertices of an isosceles triangle.

5. Arrange[81] 8 points in space, so that all 56 triangles (C_8^3) whose vertices are in the given points are isosceles triangles.

Study the problem of distributing n points in a drawing ($n = 7, 8, 9, \ldots$) so that among the C_n^3 triangles there were as many as possible isosceles ones. An analogous question may be set (for $n = 9, 10, 11, \ldots$) for space.

All these problems may be considered also in m-dimensional space, if we regard points as being sets of m real numbers written down is a definite order, and the distance between the points

$$A(\alpha_1, \alpha_2, \ldots, \alpha_m) \text{ and } B(\beta_1, \beta_2, \ldots, \beta_m)$$

is calculated from the formula

$$d = \sqrt{(\alpha_1-\beta_1)^2+(\alpha_2-\beta_2)^2+ \ldots +(\alpha_m-\beta_m)^2}.$$

6. Using the latter formula, show[82] that in four-dimensional space, every 3 of the following 5 points

$$O\,(0, 0, 0, 0), \quad A(1, 0, 0, 0), \quad B\left(\frac{1}{2}, \frac{\sqrt{3}}{2}, 0, 0\right),$$

$$C\left(\frac{1}{2}, \frac{\sqrt{3}}{6}, \frac{\sqrt{6}}{3}, 0\right) \quad \text{and} \quad D\left(\frac{1}{2}, \frac{\sqrt{3}}{6}, \frac{\sqrt{6}}{12}, \frac{\sqrt{10}}{4}\right),$$

are the vertices of an equilateral triangle.

Problems 2 and 3 are close to an interesting problem in projection geometry — the construction of configurations.

A *plane configuration* (p_m, q_n) is the name of a system of p points and q straight lines, in which m straight lines pass through each point and each straight line passes through n points (of the given ones).

It is easy to prove[83] that $pm = qn$ must take place. If $p = q$ and $m = n$, then instead of (p_m, q_n) we can write (p_m) for short.

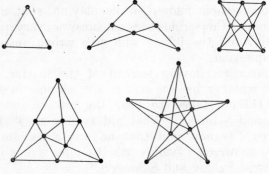

Fig. 121.

Figure 121 shows configurations (3_2), $(6_2, 4_3)$, (9_3), (9_3) and (10_3). It is interesting that there does not exist, say, the configuration (7_3) (see [8], pp. 107–109).

§. 36. Problems of a Logical Nature

Problems of a logical nature deserve attention not only because of their apparent entertainment value, but also, because in solving them, as in solving mathematical problems, we acquire quickness of mind, persistence, and the ability of finding a weak spot in the conditions of the problem, of which we can take advantage in finally solving it. Let us quote several examples:

1. Some schoolchildren, while playing, split up into two groups: the serious ones who answer every question correctly, and the jokers who give wrong answers to every question.

The teacher, having learned of the matter, asked Ivanov whether he was one of the serious ones, or a joker. Having failed to hear the answer, he asked Petrov and Sidorov; "What did Ivanov say?" Petrov answered "Ivanov said that he was a serious one". Sidorov answered "Ivanov said that he was a joker". What were Petrov and Sidorov?([84]).

2. 6 schoolchildren, who were taking part in Sunday work (Translator's note: there is a Soviet custom of asking schoolchildren or employees of an institution to put in an occasional Sunday's voluntary work on some project) split up into 3 teams. The leaders of the teams were Volodya, Petya, and Vaysa; Volodya and Misha were given logs 2 metres long; Petya and Kostya were given logs $1\frac{1}{2}$ metres long and Vaysa and Alyosha were given logs 1 metre long. The logs were sawn into $\frac{1}{2}$-metre lengths.

222

The wall newspaper reported that the team leader Lavrov with Rozhkov have sawn up 26 lengths, the team leader Galkin with Komkov have sawn up 27 lengths and team leader Kozlov with

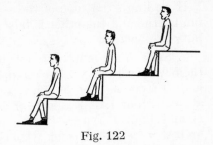

Fig. 122

Yevdokimov have sawn up 28 lengths of log. What is Komkov's first name?[85].

3. Three friends, Andrey, Boris and Vadim were sitting bareheaded one behind the other (Fig. 122). Boris and Vadim were not allowed to look behind them. Boris could see the head of Vadim, who was sitting below him, and Andrey could see the heads of both his friends.

Each friend had a hat of a colour unknown to him, placed on his head. It was taken out of a sack containing two white and three black hats (this was known to all three of them). Two hats of a colour unknown to all remained in the sack.

Andrey announced that he couldn't determine the colour of his hat. Boris heard his friend's answer and said that he, too, lacked data to determine the colour of his hat. Could Vadim determine the colour of his hat on the basis of his friends' answers?[86].

Of particular note are problems, whose solutions require a meticulous analysis of numerous data, at first glance little connected with the quantities required. For example:

4. Five schoolchildren took part in a bicycle race. After the race five fans announced:

    I. Seryozha was second and Kolya was third.
   II. Nadya was third and Tolya was fifth.
  III. Tolya was first and Nadya was second.
  IV. Seryozha was second and Vanya was fourth.
   V. Kolya was first and Vanya was fourth.

If we know that one of the statements made by each fan is true and the other is false, find the correct distribution of places([87]).

5. Sixteen students were returning to Leningrad after their winter holidays. It so happened that four, A, B, C and D were natives of Kiev, four, E, F, G, and H of Moscow, four I, J, K and L of Saratov, and four, M, N, O and P of Fergana. It also happened that A, E, I and M were 20 years old; B, F, J and N were 21; C, G, K and O were 22; D, H, L and P were 23.

Among them there were four mathematicians, four chemists, four geologists and four biologists, and any four students of one subject hailed from different towns and were of different ages.

Four students attended 1st year classes, four were second year students, four were third year and four were fourth year students, and any four attending classes of the same year came from different towns, were of different ages and studied different subjects.

Finally, four were enthusiastic footballers, four were boxers, four were volleyball players, and four were chess players, and followers of any one of the sports came from different towns, were of different ages, studied different subjects and attended classes of a different year.

Establish the subject, the year of study and the favourite sport of each student, if it is known, that *I* is a volleyball player, *F* is a footballer, *C* is a biologist, *D* is a mathematician in his 1st year of study and a chess player, *G* is a chemist in his 2nd year of study and a chess player and *J* is a geologist in his third year of study and a chess player.

For greater clarity, it is convenient to construct a table (see below).

The compartments of this table should contain the department, year and favourite sport of each student.

The table shows the information given in the problem; it remains to fill([88]) the blanks in the table.

Age Town	20 years	21 years	22 years	23 years
Kiev	A, —,—,—.	B, —,—,—.	C, biol,—.—	D, math. I, chess
Moscow	E,—,—,—.	F,—,—, footb.	G, chem, II, chess	H,—,—,—.
Saratov	I,—,—. volleyb.	J, geol. III. chess	K,—,—,—.	L,—,—,—.
Fergana	M,—,—,—.	N,—,—,—.	O,—,—,—.	P,—,—,—.

Logical problems include those concerned with rediscovering erased figures, and with the deciphering of arithmetical operations in which figures are replaced by letters.

Solving problems of this kind develops logical thinking, and the invention of new problems of this type can serve as a source of entertainment.

In the examples quoted below, it is possible, by taking all circumstances into account, to reinstate the erased numbers, and to decode the values of the letters (uniquely in each example) ([89]).

I. Reinstate the erased numbers in the operations written down below

a)
$$\sqrt{******} = ***$$
$$\underline{\quad\ *}$$
$$\overline{\quad ***}$$
$$\overline{\,-\,**}$$
$$\overline{\quad 4***}$$
$$\overline{\,-\,****}$$
$$\overline{\qquad 0}$$

b)
$$****** \quad ***$$
$$\underline{\,-\,****} \quad *8*$$
$$\overline{\quad ***}$$
$$\overline{\,-\,***}$$
$$\overline{\quad ****}$$
$$\overline{\,-\,****}$$
$$\overline{\qquad 0}$$

II. Find the values of the letters in the following notation of summing of multi-digit numbers:

(q)
   s m e h
   g r o m
   ―――――
   g r e m i

(b)
   forty
   ten
   ten
   ―――――
   sixty

(similar letters must be replaced by the same digit throughout an example, and different letters by different digits.)

These problems are related to the so-called arithmetical rebuses, in which the operations of dividing two multidigit numbers can be written down in code, by taking some ten-letter word as a "key" and replacing mentally each successive letter of the word by the digits 1, 2, 3, 4, 5, 6, 7, 8, 9, 10 respectively. It is best to set the rebus to a group of several persons, gradually writing down (at intervals of 2 to 3 minutes) the successive digits of the quotient, of the product of these digits and the divisor, and of the corresponding remainders, until one of the solvers finds the "key" of the rebus.

Suppose the word "troodolubye" is to serve as a "key" and the operation to be coded is that of dividing 240176 by 119. Having carried out the division in numbers, for convenience, the "leader" then writes it down in the coded form;

$$(1) \left\{ \begin{array}{c} \overline{r\ d\ e\ t\ u\ l} \\ \overline{\left\lceil r(oo)b \right. } \\ \overline{rtu} \end{array} \right. \qquad \begin{array}{c} \left| t\ t\ y \right. \\ \overline{r\ e\ t} \\ \overset{\smile}{(1)}\ \overset{\smile}{(2)} \end{array}$$

$$(2) \left\{ \begin{array}{c} \overline{\left\lceil ttu \right.} \\ \overline{ybl} \end{array} \right.$$

First, the letters marked (1) should be written down; here the solvers can notice already that from $r \times tty = r(oo)b$ it follows that $r \times m \leqslant r$ and, therefore, $m = l$: it may be also guessed that the next letter in the quotient has the value zero.

After the letters marked (2) have been written down, the solvers may notice that $r = 2$ (*rtu − tty* is a two-digit number, therefore $r - t = 1$) and $b = 8$ (since $e = 0$ and $r(oo)b + r = r\ d\ e$).

If someone, on writing these letters into his spaces for the "key" required, guesses that the "key" is the word "troodolubye", then the following portion of letters written down may, generally speaking, either confirm or upset his guess (in this case, it will confirm it).

The complexity of problems of this type increases considerably, if the digits are coded by ten different letters, not forming any "word-key".

We shall quote three more problems in whose solutions the logical element predominates. The last one can also serve as an interesting illustration of the theory of indeterminate equations of the first degree.

Difficult Crossings

Three merchants A, B and C and their servants a, b and c must cross a river in a two-seater boat in such a way that no servant remains without his master in the company of even one of the remaining merchants.

One of the possible solutions is given by the scheme

crossing	ab	bc	AB	BC	ab	ac
returning	b	c	Bb	a	a	

If n merchants with servants are taken, then, for $n = 4$ and $n = 5$, the problem becomes insoluble for a two-seater boat, but has a solution if a three-seater boat is used.

We suggest that the reader shows how the crossing is realized

(a) when $n = 2$ (two-seater boat), by making three "direct" journeys.

(b) when $n = 4$ (three-seater boat) by making five "direct" journeys,

(c) when $n = 5$ (three-seater boat) by making six "direct" journeys ([90]).

When $n = 6$, a three-seater boat is no longer adequate, but, evidently, for any value n, a four-seater boat is sufficient.

If an island, where temporary landings may be made, is introduced, then the problem can be solved with a two-seater boat, for any value of n.

The Detection of a False Coin

Lately, problems on detecting false coins, differing from the normal ones by weight only, have become very popular.

In the simplest form of the problem, it is required to detect a single (lighter) coin in a pile containing $n = 3^k$ coins, by means of k weighings on a beam balance (without weights).

Here it is sufficient to divide the coins into three groups with 3^{k-1} coins in each, and to place any two of these on the pans; this shows the group of 3^{k-1} coins containing the false coin immediately.

Proceeding in the same way with this group, we find the group of 3^{k-2} coins containing the false coin, etc. Problems in which it is not stated whether the coin is lighter or heavier, are more complex.

Here is one of the possible solutions of the following problem:

One of twelve coins is false. Detect this coin by means of three weighings and determine whether it is heavier or lighter than the normal coin (see also [29] part I problem 6).

If at each weighing we put four coins on each pan of the balance, as shown in the second and third columns of the table below (only the numbers ascribed to the coins are shown) it is easy to see that for 24 possible assumptions:

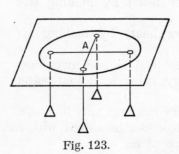

Fig. 123.

the false coin is light and its number is 1, 2, 3, 4, 5, 6, 7, 8, 9, 10, 11, 12, and: *the false coin is heavy and its number is* 1, 2, 3, 4, 5, 6, 7, 8, 9, 10, 11, 12 (see the lower rows of the table) we obtain 24 different possibilities of the results of weighing, shown in the numbered columns of the table.

Possible results of weighings — in columns.

Notation: l — lighter on the left, t — heavier on the left,
r — equal weight on both right and left.

No. of weighing	coins on the left	coins on the right	1	2	3	4	5	6	7	8	9	10	11	12	13	14	15	16	17	18	19	20	21	22	23	24
1.	1, 2, 3, 4	5, 6, 7, 8	l	l	l	l	t	t	t	t	r	r	r	r	t	t	t	t	l	l	l	l	r	r	r	r
2.	1, 2, 3, 5	4, 9, 10, 11	l	l	l	t	l	r	r	r	t	t	t	r	t	t	t	l	t	r	r	r	l	l	l	r
3.	1, 6, 9, 12	2, 5, 7, 10	l	t	r	r	t	l	t	r	l	t	r	l	t	l	r	r	l	t	l	r	t	l	r	t

No. of the false coin		1	2	3	4	5	6	7	8	9	10	11	12	13	14	15	16	17	18	19	20	21	22	23	24
	Light:	1	2	3	4	5	6	7	8	9	10	11	12												
	Heavy:													1	2	3	4	5	6	7	8	9	10	11	12

16   229

Having established, in the course of the weighings, which situation actually takes place, we find in the two lowest rows of the table the answer to the questions posed in the problem.

Problems of this kind are being generalized in various ways (see, for example, [29], part I, problem 5), and give an example of a mathematical pastime awaiting the construction of a complete theory.

For the sake of variety, the use of a balance with m pans can be admitted (Fig. 123, where $m = 4$), enabling us to find, at one weighing (from the position of the nodal point A), which of the m groups of coins is lighter (or heavier) than the rest.

Problems on Dividing Liquids

We first consider a problem encountered (in a slightly different form) in Chuquet (1484) and in Tartaglia (1586).

It is required to pour off 4 litres of wine from a full vessel of capacity 8 litres, using two empty vessels of capacity five and three litres respectively.

Suppose that we first pour the liquid into the medium vessel. If we avoid the obviously superfluous actions, it is easy to arrive at a solution represented by the scheme:

$$8, 0, 0 \rightarrow 3, 5, 0 \rightarrow 3, 2, 3 \rightarrow, 6, 2, 0 \rightarrow 6, 0, 2 \rightarrow$$
$$\rightarrow 1, 5, 2 \rightarrow 1, 4, 3 \rightarrow 4, 4, 0.$$

It can be seen from this scheme that the liquid is poured each time into the *empty medium vessel (filling it)* and then it is *returned in portions equal to the capacity of the small vessel.*

In another method of solution of the problem, the small and the medium vessels exchange their roles: the liquid is poured from the large vessel into the *empty small vessel (filling it)* and it is *returned in portions equal to the capacity of the medium vessel*:

$$8, 0, 0 \rightarrow 5, 0, 3 \rightarrow 5, 3, 0 \rightarrow 2, 3, 3 \rightarrow 2, 5, 1 \rightarrow$$
$$\rightarrow 7, 0, 1 \rightarrow 7, 1, 0 \rightarrow 4, 1, 3 \rightarrow 4, 4, 0.$$

In general, if we denote the capacities of three vessels by a, b, c (a is an even number, $a > b > c$, and, of course, $b + c \leq \frac{1}{2} a$), then, if b and c are relatively prime and if $a \leq b + c - 1$, both methods lead to the result which follows from the solubility in positive integers of the equations

$$a - bx + cy = \frac{a}{2} \quad \text{and} \quad a - cu + bv = \frac{a}{2},$$

corresponding to the first and second methods respectively (see Ch. 4).

But, as soon as $a = b + c - 2$, one of the methods may turn out to be useless (for instance, if, in the first method, there arises the situation $(b - 1, 0, c - 1)$), it is impossible to fill the medium vessel from the large one and it is impossible to pour c litres of wine from the small vessel into the large one, but then the second method must give the result (see [25], ch. 3). For example, when $a = 20$, $b = 13$ and $c = 9$, ten litres of wine can be separated out only by the second method[91].

For $a < b + c - 2$, the problem may turn out to be insoluble, which the reader may verify by taking, for instance, $a = 16$, $b = 12$, and $c = 7$[92].

§ 37. Rag-Bag

In this chapter problems are collected from various branches of mathematics. Some of them are accessible to a wide circle of readers, others are intended for persons with mathematical training. In a number of cases, a pencil, paper and persistence are all that is required. Certain problems may suggest themes for independent research.

The Geometrical Demonstration of the Formula

$$1^2+2^2+3^2+ \ldots +(n-1)^2+n^2 = \frac{n(n+1)(2n+1)}{6}$$

Suppose cubes of unit edge are arranged in n layers, there being n^2, $(n-1)^2, \ldots, 3^2, 2^2, 1^2$ cubes in successive layers from bottom to top, (in Fig. 124, $n = 5$). On circumscribing the pyramid $OADBC$, which has $OA = OB = OC = n + 1$ (its volume equals $\frac{(n+1)^3}{3}$ about the whole pile of cubes, we find, that the sum of volumes of all cubes equals

$$1^2+2^2+3^2+ \ldots +(n-1)^2+n^2 =$$
$$= \frac{(n+1)^3}{3} -[n+(n-1)+ \ldots +3+2+1] -$$
$$-(n+1) \cdot \frac{1}{3} = \frac{(n+1)^3}{3} - \frac{n(n+1)}{2} - \frac{n+1}{3} =$$
$$= \frac{n(n+1)(2n+1)}{6} ;$$

ere the expression in the square brackets equals the

sum of the volumes of prisms $AFHEPQ$ $\left(\text{volume} = \dfrac{n}{2}\right)$ $KPQBML$ $\left(\text{volume} = \dfrac{n}{2}\right)$ and similar prisms in all the remaining layers: $\dfrac{1}{3}(n+1)$ is the sum of volumes of little pyramids, ($EPKDQ$, etc.).

Try to prove in a similar way that

$$1^2 + 3^2 + 5^2 + 7^2 + \ldots + (2n-3)^2 + (2n-1)^2 = \frac{n(4n^2-1)}{3} \quad (93).$$

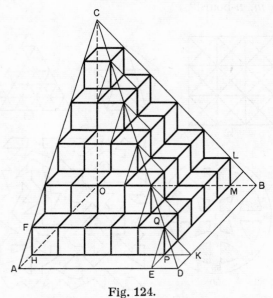

Fig. 124.

Exercises for the Development of Geometrical Intuition

Below, several problems are given, each of which can be used in competitions with the motto "who can spot most".

To solve problems of this kind, we have to be systematic in counting up the figures of the given type discovered in the drawing; in the problem No. 4 it is necessary to think out meticulously in what order to pick out the figures of the form given so that none of them is missed.

The reader can invent any number of similar problems without any difficulty.

1. How many triangles, squares and rectangles can be seen in Fig. 125a?([94]).

2. Determine the number of triangles in Fig. 125b([95]).

3. How many triangles, regular hexagons and rhombi are there in Fig. 125c?([96]).

4. How many squares and how many rectangles can be seen in an ordinary chessboard?; in an n^2-board?; in an m, n-board?

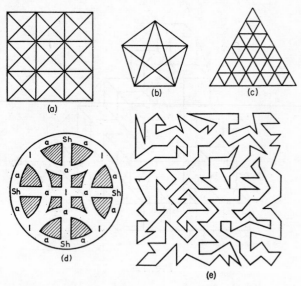

Fig. 125.

How many cubes and how many rectangular parallelepipeds can be seen in a cube of edge 10 cm, which is subdivided by planes parallel to the faces of the cube into little cubes of edge 1 cm?([97])

5. In how many ways can the word (sh)ala(sh) be read, by moving along straight paths, curved paths and broken paths (Fig. 125d) if the initial and the final letters (sh) are not to coincide?([98])

Let us also mention a game which can be conducted with any polygon bounded by a very complicated non-convex line (see, e. g. Fig. 125*e*): a point is placed in the drawing, and it is required to tell as quickly as possible, whether the point is situated inside or outside the polygon.

Interesting Identities

1. It is easy to verify, that $(3s^{2n} - 2s^n - 1)^2 + 4s^{2n} + 4s^n)^2 \equiv (5s^{2n} + 2s^n + 1)^2$. Therefore, from the formulae

$$a_n = 3s^{2n} - 2s^n - 1: \quad b_n = 4s^{2n} + 4s^n: \quad c_n = 5s^{2n} + 2s^n + 1$$

(where s and n are natural numbers and $s > 1$) it is possible to obtain any number of pythagorean triples of numbers (see § 5).

The angles of a right-angled triangle of sides a_n, b_n, c_n, when n is sufficiently great, differ by as little as we please from those of a triangle with sides 3, 4, 5, for

$$\lim_{n \to \infty} \frac{a_n}{b_n} = \lim_{n \to \infty} \frac{3s^{2n} - 2s^n - 1}{4s^{2n} + 4s^n} = \lim_{n \to \infty} \frac{3 - \dfrac{2}{s^n} - \dfrac{1}{s^{2n}}}{4 + \dfrac{4}{s^n}} = \frac{3}{4} .$$

Try to find an analogous identity, leading to rectangular triangles with integral sides "nearly similar" to a triangle of side 5, 12, 13.

2. Look for identities in which the product of two "unwieldy" polynomials equals a polynomial with few terms; for example $(x^8 - 4x^7 + 8x^6 -$

$$-10x^5 + 8x^4 - 4x^3 + 2x^2 - x + \frac{1}{4}\Big)\Big(x^8 + 4x^7 + 8x^6 + 10x^5 +$$

$$8x^4 + 4x^3 + 2x^2 + x + \frac{1}{4}\Big) \equiv x^{16} + \frac{17}{2}x^8 + \frac{1}{16} .$$

3. The identity $\dfrac{a^3 + b^3}{a^3 + (a-b)^3} \equiv \dfrac{a+b}{a + (a-b)}$ can be used for "inadmissible cancellations" leading to correct results, for example

$$\frac{37^3+13^3}{37^3+24^3} = \frac{37+13}{37+24} = \frac{50}{61}$$

Perhaps you could succeed in finding similar identities, where the indices of the fourth power could be "cancelled" in the numerator and the denominator.

4. The identity $\dfrac{\log\left(\dfrac{m+1}{m}\right)^m}{\log\left(\dfrac{m+1}{m}\right)^{m+1}} \equiv \dfrac{\left(\dfrac{m+1}{m}\right)}{\left(\dfrac{m+1}{m}\right)^{m+1}}$ (m is a positive rational number) shows that sometimes the inadmissible cancellation of the log sign: $\dfrac{\log a}{\log b} = \dfrac{a}{b}$ can lead to a correct result. For example

$$\frac{\log\dfrac{9}{4}}{\log\dfrac{27}{8}} = \frac{\dfrac{9}{4}}{\dfrac{27}{8}} \text{ (here } m=2).$$

5. The identity $\sqrt[n]{a + \dfrac{a}{a^n - 1}} \equiv a\sqrt[n]{\dfrac{a}{a^n - 1}}$ leads to a number of curious equations. For example:

$$\sqrt[3]{2\tfrac{2}{7}} = 2\sqrt[3]{\tfrac{2}{7}}; \quad \sqrt[4]{5\tfrac{5}{624}} = 5\sqrt[4]{\tfrac{5}{624}}; \quad \sqrt[5]{2\tfrac{2}{31}} = 2\sqrt[5]{\tfrac{2}{31}} \text{ etc}$$

6. By cancelling, quite without reason, the symbol sin in the numerator and in the denominator of the right-hand-side of the identity

$$\sin \alpha + \sin 2\alpha + \ldots + \sin n\alpha \equiv \frac{\sin\dfrac{(n+1)\alpha}{2} \cdot \sin\dfrac{n\alpha}{2}}{\sin\dfrac{\alpha}{2}}$$

and then in both sides of the identity, we arrive at the identity

$$\alpha + 2\alpha + \ldots + n\alpha \equiv \frac{\dfrac{n+1}{2}\alpha \cdot \dfrac{n}{2}\alpha}{\dfrac{\alpha}{2}}$$

Verify this identity.

7. Prove ([100]), that $[3(10^k + 10^{k-1} + \ldots + 10^2 + 10 + 1)n + 1]^2 \equiv n^2(10^{2k+1} + 10^{2k} + \ldots + 10^{k+1}) + (6n - n^2)(10^k + 10^{k-1} + \ldots + 10 + 1) + 1$.

This identity leads to two series of interesting numerical equations: for $n = 1$ and $k = 1, 2, 3, 4, \ldots$ we have $34^2 = 1156$; $334^2 = 111556$; $3334^2 = 11115556$; $33334^2 = 1111155556$, etc., for $n = 2$ and $k = 1, 2, 3, 4, \ldots$ also we have: $67^2 = 4489$; $667^2 = 444889$; $6667^2 = 44448889$; $66667^2 = 4444488889$, etc.

Try to find identities leading to similar series of numerical equations in other scales of notation.

Optical Illusions

Try to look at a distant wall through your index fingers, whose tips you have made to touch about

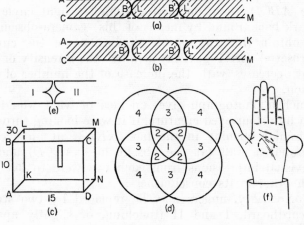

Fig. 126.

35–50 cm. away in front of your face. You will have the impression that your fingers keep a little "sausage" between them, which remains "suspended" in the air when you pull your fingers apart slightly.

The length of this "sausage" is the greater the farther is the object viewed "through the fingers". The number of "sausages" may be increased by making several fingers of your hands touch in pairs.

This strange phenomenon can be explained very simply: the portion of wall enclosed by the lines ABC and KLM cannot be seen by the right eye (Fig. 126a, b), and the portion of wall enclosed by the lines $AB'C$ and $KL'M$ cannot be seen by the left eye. In total the portion of wall, which cannot be seen altogether has the shape of the figure shaded-in in Fig. 126a, b.

For the same reason a narrow vertical slit cut out of the face $KLMN$ of a box of approximate dimensions $10 \times 15 \times 30$ cm³ (see Fig. 126c) is seen as two parallel slits (or as a wide slit with a vertical join), if we look through the slit with both eyes at a wall facing the "absent face" $ABCD$.

If we replace the slit in the face $KLMN$ by four point-apertures distributed at the vertices of some square, and if we place a photographic plate in the position of face $ABCD$, then, on photographing a light circle on a dark background by means of this "camera-obscura", we obtain (for appropriate dimensions of the circle) a "rosette" (see Fig. 126d) in which the intensity of the light decreases with the increase of the number of the region.

Such a photograph gives an idea of what would be seen by a four-eyed creature, if it were looking through a circular opening in the face $NKLM$ at an evenly illuminated wall (at the absent face $ABCD$) and if its visual impressions from each eye were to overlap each other in its consciousness.

For variety, fingers may be replaced by two strips of cardboard, I and II (touching, or slightly apart) of this or that profile (Fig. 126e); parallel vertical slits may be replaced by curved slits. By photographing various objects and by selecting the point-apertures in the "camera-obscura" in various ways, amateur photographers may obtain unusual "geometric photographs".

In conclusion, we shall mention one curious optical illusion: if we look with our right eye through a tube at some object and if we screen it with the left hand, which touches the tube, from the left eye, we get the impression, that the object can be seen also by the left eye, through a "hole in the hand" (Fig. 126*j*).

Miscellaneous Problems

First, we shall quote several simple problems and questions.

1. One tumbler contains m cm³ of water and another tumbler contains n cm³ of spirits. First, a cm³ of water were poured from the first tumbler into the second, then (after thorough mixing) a cm³ of the mixture were poured back.

If it is assumed, for the sake of simplicity, that the volume of the mixture equals the sum of volumes of the liquids involved, determine, whether the amount of spirits mixed with the water in the first tumbler (by volume) is greater or less than the amount of water in the second tumbler?[101].

2. Petrov was asked "whose portrait is hanging on the wall?" Petrov answered "The father of the one that is hanging is the only son of the father that is speaking". Whose portrait was it?[102].

3. How many great-great-grandparents altogether had all your own great-great-grandparents?[103].

4. How would an angle of 15′ look, when you look at it through a magnifying glass of magnification 4?[104]

5. By how many per cent does the purchasing power of the population increase, if the prices of all goods drop by 20%?[105].

6. If the purchasing power of the population increased first by 20%, then by 25%, by how many percent did the purchasing power of the population increase altogether?[105a].

7. How many times do the hour and minute hands of a clock form a right angle in 24 hours?[106].

8. When Kolya was as young as Olya is now
The years of Aunt Polya
Were as many as now
Are those of Kolya together with
Those of Olya.
How old was Kolya
When Auntie Polya
Was the age of Kolya?[107].

9. A pilot flew from point A and, having flown 800 km due south (to the point B), he then flew due east. On flying 800 km more (to the point C) he noticed a bear underneath. What colour was the bear, given that $AB = AC$?[108]

10. A and B sold a herd of oxen, and obtained as many roubles for each ox as there were oxen in the herd. Desiring to divide the money equally, they each took 10 roubles in turn from the sum obtained. A got 10 roubles extra and added his purse to the remainder in compensation. What was the price of the purse?[109].

11. Three men with their wives entered a shop. Each of the six persons bought several articles, and paid for each article as many roubles as the number of articles he or she bought.

Each husband spent 45 roubles more than his wife; Yuri spent 525 roubles more than Olga, Login spent 13 roubles more than Nina. The names of the rest were Alexander and Tatyana.

Who is married to whom and how many articles did each person buy?[110].

12. Someone divided a sum of money in his possession among some children. The first child received one rouble and $\frac{1}{6}$ of the remainder, the second received two roubles

and $\frac{1}{6}$ of the new remainder, the third received three roubles and $\frac{1}{6}$ of the new remainder, etc.

Given that all the money (S roubles) was distributed equally among the children, determine S and the number of children.

It is remarkable that by equalizing the shares of the children we get a set of *simultaneous* equations with S unknown, giving $S = 25$. The square of any natural number possesses an analogous porperty[111]: taking out of n^2, 1 and $\frac{1}{n+1}$ of the remainder, then 2 and $\frac{1}{n+1}$ of the new remainder, etc., we obtain a series of equal amounts.

Perhaps the reader might find some other class of numbers whose properties could serve as a source of a series of entertaining problems?

13. Pavel walked from M to get to N. At exactly the same time Gleb left M on a motorcycle, whose driver was Yuri. After travelling part of the way, Gleb continued on foot, while Yuri drove back to meet Pavel and took him on as a passenger. All arrived at N simultaneously. Knowing that $MN = s$ km, the speed of a man on foot is u km/hr and the speed of a motorcycle is v km/hr, find the time taken by the three friends to get from M to N[112].

This problem can be generalized by supposing, for example, that the motorcyclist (or two motorcyclists, travelling at speeds of v_1 km/hr and v_2 km/hr) helps a group of n friends to arrive at their destination simultaneously, it being also supposed that one of the motorcyclists aiding the walkers can transport two passengers at a time, etc.

Try to investigate various variants of the problem, making up timetables of movements of the group on every occasion.

14. A certain job was started after 4 o'clock and finished after 7 o'clock, and the time shown on the

clock at the beginning of the job is changed into the time shown at the end of it by interchanging the positions of the hour hand and the minute hand.

Determine the duration of the job and show that the hands are equally inclined to the vertical at the beginning and at the end of work.([113])

15. How many times in 24 hr do the times on the clock have the property, that the interchange of the hour hand and the minute hand leads to meaningful readings of the clock?([114]).

16. Prove that for every natural number k the total number of digits in the sequence of numbers 1, 2, 3, ... $10^k - 1$, 10_k equals the number of zeros in the sequence of numbers 1, 2, 3, ... $10^{k+1} - 1$, 10^{k+1}([115]).

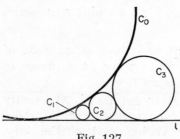

Fig. 127

17. Prove that a plane figure S of arbitrary form, whose area is less than one square centimetre can be laid out on squared paper with squares 1 cm² in such a way that it does not cover any of the nodal points.([116])

18. Is it possible to draw a straight line to intersect given straight lines l_1, l_2, l_3, l_4, which are arranged in space in a random fashion?

19. A number of circles C_1, C_2, C_3, ... are drawn between a straight line l and a circle C_0 touching it, in such a way that C_{k+1} touches the circles C_k and C_0 and the straight line l.

Find the radius of the circle C_{1000}, if the radius of circle C_0 equals 1 km and the radius of circle C_1 is 1 mm.([118])

20. Prove that a ray of light, having been reflected from three mutually perpendicular mirrors in turn becomes parallel to its original direction but in the opposite sense.([119])

21. How should a ray of light be directed inside a rectangular parallelepiped with mirror faces, in order that it returns to its starting point after being reflected from all six faces?[120]

22. A pilot flew 2000 km due south, 2000 km due east and 2000 km due north. He arrived at his starting point. Where did he fly from? (The problem has many solutions).[121]

23. An aeroplane flew out of Leningrad, and having travelled a km due north, a km due east, and a km due south found himself $3a$ km east of Leningrad. Find a[122].

24. What is the area of an equilateral triangle drawn on the surface of the earth, if each of its angles equals 72°? Here we are concerned with a so-called *spherical triangle*, formed by the arcs of great circles drawn through the vertices of the triangles[123].

25. Determine the angles of an equilateral triangle of side 1 km drawn on the perfectly smooth surface of a frozen lake[124].

Puzzles

1. Show, that it is possible to pass from the "pleating" (*a*) in Fig. 128, to the pleating (*b*) without breaking the string rings (see [12] for analogous problems).

2. How should three string rings be linked, so that by cutting one of them, the other two could be separated without additional cutting?

Solve the analogous problem for n string rings[125].

3. If we treat the meaning of the word "inside" with a slight degree of freedom, we can solve the following jocular problem: find three completely identical bodies, *A*, *B* and *C*, which can be arranged in such a way that *A* is inside *B*, *B* is inside *C*, *C* is inside *A*.

The conditions of the problem are satisfied by slightly stretched frames, "fitted into each other" — see Fig. 128c.

A similar problem can be solved with respect to n identical solids. Let n snakes (they can be imagined in the form of narrow thin walled "conical" sacks) be situated along a circle of radius R and each begin to swallow the snake in front of it with equal "speeds of swallowing" for all snakes (see Fig. 129a, where $n = 2$).

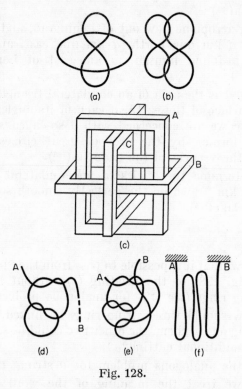

Fig. 128.

When each snake swallows half of its victim, there appears a double ring of radius $\frac{1}{2}R$ (Fig. 129b); when each of the snakes is 99% inside the snake swallowing it, the ring of radius $R/100$ consists of one hundred layers, etc.

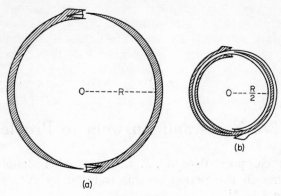

Fig. 129.

4. After tying a piece of string in the form of the knot (*d*) (Fig. 128) and after passing from it to the knot (*e*) it is easy to show, that on pulling ends *A* and *B*, no knot remains in the string.

It follows that when the ends *A* and *B* are fixed, the string can be folded in the form of the "knot" (*e*). Try to carry this out!

§ 38. Notes and Answers to Problems

(1) Suppose $0 \cdot (a_1 a_2 \ldots a_{s-1} a_s)_{(k)} = \alpha$; denote the number in the period of this fraction by N: $a_1 a_2 \ldots a_{s-1} a_{s(k)} = N$,
then

$$\alpha = \frac{a_1}{k} + \frac{a_2}{k^2} + \ldots + \frac{a_{s-1}}{k^{s-1}} + \frac{a_s}{k^s} + \frac{a_1}{k^{s+1}} + \frac{a_2}{k^{s+2}} + \ldots$$

$$\ldots + \frac{a_{s-1}}{k^{2s-1}} + \frac{a_s}{k^{2s}} + \ldots = \frac{a_1 k^{s-1} + a_2 k^{s-2} + \ldots + a_s}{k^s} +$$

$$+ \frac{a_1 k^{s-1} + a_1 k^{s-2} + \ldots + a_s}{k^{2s}} + \ldots = \frac{N}{k^s} + \frac{N}{k^{2s}} + \frac{N}{k^{3s}} + = $$

$$= \frac{N}{k^s - 1} = \frac{N}{\overline{k-1}\ \overline{k-1} \ldots \overline{k-1}\ \overline{k-1}_{(k)}}$$

(the s-digit number in the denominator is written down by means of s digits "$k - 1$").

(2) Suppose the square root is extracted from the four-digit number $N = abcd_{(k)}$ (a, b, c, d are the digits of number N as written down in the system of notation with base k; below we shall omit to add the suffix "k". If the first digit α of the root required is found, i. e. $\alpha^2 \leq ab = ak + b < (\alpha + 1)^2$, then

$$\underline{\sqrt{abcd}} = \alpha \ldots$$
$$\frac{\alpha^2}{a'b'cd}$$
$$\ldots \ldots \ldots$$

Evidently, $N = abcd = \alpha^2 k^2 + a'b'cd$.

Now, it is necessary to find the greatest digit, β, for which $(\alpha k + \beta)^2 = \alpha^2 k^2 + 2\alpha \beta k + \beta^2 \leq N$, or $(2\alpha k + \beta)\beta \leq N - \alpha^2 k^2 = a'b'cd$, i. e. we seek the greatest digit β, which, when written down on the right of the number 2α, gives a number (equal to $2\alpha k + \beta$), whose product with β does not exceed the remainder $a'b'cd$. But the same rule is followed in extracting the square root in the decimal system of notation.

In extracting roots from numbers which can be subdivided into three divisions, we denote by α the two-digit number, which is obtained, in the manner indicated, in the extraction of the root from the two senior divisions, and by β, the required units digit, all the reasoning given above remains valid.

For numbers which can be split up into four divisions, the root of the first three senior divisions is sought first, and then — by the method indicated — the units digit of the required root, etc.

(^{3a}) $2713 = 41323_{(5)} = 41332_{(5)} = 414\overline{2}\overline{2}_{(5)} = 1\overline{1}2\overline{1}\overline{2}\overline{2}_{(5)}$.

Check: $1\overline{1}2\overline{1}\overline{2}\overline{2}_{(5)} = 5^5 - 5^4 + 2 \cdot 5^3 - 5^2 - 2 \cdot 5 - 2 =$

$= 2713;\ 409 = 3114_{(5)} = 12\overline{1}2\overline{1}_{(5)}$.

($3b$) Let $N = abc_{(8)} = a \times 8^2 + b \times 8 + c$ where none of the digits a, b and c exceeds seven. If $a = \alpha_1\alpha_2\alpha_{3(2)}$, $b = \beta_1\beta_2\beta_{3(2)}$, $c = \gamma_1\gamma_2\gamma_{3(2)}$, where each digit α_1, α_2, α_3, β_1, β_2, β_3, γ_1, γ_2, γ_3 equals either zero or unity, then

$$N = (\alpha_1 \cdot 2^2 + \alpha_2 \cdot 2 + \alpha_3) \cdot 2^6 + (\beta_1 \cdot 2^2 + \beta_2 \cdot 2 + \beta_3) \cdot 2^3$$
$$+ (\gamma_1 \cdot 2^2 + \gamma_2 \cdot 2 + \gamma_3) = \alpha_1 \cdot 2^8 + \alpha_2 \cdot 2^7 + \alpha_3 \cdot 2^6 +$$
$$+ \beta_1 \cdot 2^5 + \beta_2 \cdot 2^4 + \beta_3 \cdot 2^3 + \gamma_1 \cdot 2^2 + \gamma_2 \cdot 2 + \gamma_3 =$$
$$= \alpha_1\alpha_2\alpha_3\beta_1\beta_2\beta_3\gamma_1\gamma_2\gamma_{3(2)}.$$

By reading these equations from right to left we obtain a rule of transition from the binary system of notation to that of base eight: three-digit numbers obtained in splitting up a number written in the binary system into divisions (from right to left) give the digits of the same number, written down in the system of notation with base eight.

(4) For any $k > 5\,123\,454\,321_{(k)} = 11111^2_{(k)}$.

(5) Since $N \leq 1000 < 1024 = 2^{10}$, the number N can be written down in the binary system, using no more than ten digits, each of which may be either zero or unity.

To determine N it is sufficient to ask ten questions:

1. Is the first digit on the right of the number N, as written down in the binary system, unity?
2. Is the second digit unity? etc.

(6) It is easy to verify, that the theorem is true for $s = 1$ and for $s = 2$. Let us apply the method of mathematical induction to prove the theorem in general: suppose the theorem to be true for $s = n$ (i. e. for numbers from 1 to $2^n - 1$ we have n cards, each of which contains 2^{n-1} numbers), prove that it is also true for $s = n + 1$.

Any number m satisfying the conditions $2^n \leq m \leq \leq 2^{n+1} - 1$ can be represented in the form $m = 2^n + x$, where $0 \leq x \leq 2^n - 1$. All these 2^n numbers find themselves in the $(n + 1)$th card with the heading 2^n ; any number m will get into any particular card only when $1 \leq x \leq 2^n - 1$ (depending on terms of form 2^a into which x breaks down); therefore in each of the preceding cards there appear 2^{n-1} new numbers (we have assumed the truth of the theorem for $s = n$). In the final count each of the $n + 1$ cards contains 2^n numbers.

(7) As is well known

$$n! = 1 \times 2 \times 3 \times 4 \times 5 \ldots (n-2) \times (n-1)\, n. \qquad (1).$$

If we examine each of these factors in turn, we find that after every p_1 "steps" we come across factors divisible by the prime number p_1; their number is $\left(\dfrac{n}{p_1}\right)$, but of those, $\left(\dfrac{n}{p_1^2}\right)$ are divisible by p_1^2, $\left(\dfrac{n}{p_1^3}\right)$ are divisible by p_1^3, etc.

Therefore the number of factors in the eqn. (1), which contain the factor p_1 exactly once, twice, three times, etc., equals the respective numbers

$$-\left[\frac{n}{p_1^2}\right], \quad \left[\frac{n}{p_1^2}\right]-\left[\frac{n}{p_1^3}\right], \quad \left[\frac{n}{p_1^3}\right]-\left[\frac{n}{p_1^4}\right] \text{ etc. Therefore}$$

$$\alpha = \left[\frac{n}{p_1}\right]-\left[\frac{n}{p_1^2}\right]+2\left\{\left[\frac{n}{p_1^2}\right]-\left[\frac{n}{p_1^3}\right]\right\}+3\left\{\left[\frac{n}{p_1^3}\right]-\right.$$
$$\left.-\left[\frac{n}{p_1^4}\right]\right\}+\ldots = \left[\frac{n}{p_1}\right]+\left[\frac{n}{p_1^2}\right]+\left[\frac{n}{p_1^3}\right]+\ldots$$

(8) $N = 2^{4561} - 2^{2280}$: here the number to be subtracted is much smaller than the diminuend. Since $\log 2^{4561} = 4561 \times 0.301029996 = 1372.997$, therefore 2^{4561} (and therefore also N) contains 1373 digits.

(9) $S(N)$ $(1+2+2^2+\ldots+2^a)\{1+(2^{a+1}-1)\} =$
$$=\frac{2^{a+1}-1}{2-1}\cdot 2^{a+1} = 2\cdot2^a(2^{a+1}-1) = 2N.$$

(10) $\left[\dfrac{1000}{7}\right]+\left[\dfrac{1000}{7^2}\right]+\left[\dfrac{1000}{7^3}\right] = 142+20+2 = 164.$

$\left[\dfrac{100}{7}\right]+\left[\dfrac{100}{7^2}\right] = 14 + 2 = 16:$

since the numerator equals $(1000!) \div (100!)$, therefore $k = 164 - 16 = 148$.

(11) If two s-digit numbers N written down side by side form a $2s$-digit number which is an exact square, this means that

(1) $10^{s-1} \leq N < 10^s$ and
(2) $(10^s + 1) \times N$ is a perfect square.

But then $10^s + 1$ must be divisible by the square of some integer (otherwise, the smallest value N satisfying the second condition is $10^s + 1$, which contradicts the first condition).

For the determination of the smallest value of s, for which $10^s + 1$ is divisible by p^2, it is convenient, using the properties of congruences (see § 3) to seek the smallest root of the congruence $10^s \equiv -1 \pmod{p^2}$; for example, for $p = 11$ we have

$$10^2 \equiv -21 \pmod{121}; \qquad 10^3 \equiv -210 \equiv 32 \pmod{121};$$
$$10^4 \equiv 320 \equiv 78 \pmod{121}; \qquad 10^5 \equiv 780 \equiv 54 \pmod{121};$$
$$10^6 \equiv 540 \equiv 56 \pmod{121}; \qquad 10^7 \equiv 560 \equiv 76 \pmod{121};$$
$$10^8 \equiv 760 \equiv 34 \pmod{121}; \qquad 10^9 \equiv 340 \equiv -23 \pmod{121};$$
$$10^{10} \equiv -230 \equiv 12 \pmod{121}; \qquad 10^{11} \equiv 120 \equiv -1 \pmod{121}.$$

Evidently, $(10^{11}+1)\left[\dfrac{10^{11}+1}{11^2}k^2\right]$ is an exact square, if k is any natural number. It is easy to check that 4 is the smallest value of k which makes the expression in the square brackets become an eleven-digit number (equal to 13 223 140 496).

By direct inspection it can be seen that the congruence of form $10^s \equiv 1 \pmod{k^2}$, where k is any prime number, is not satisfied when $s < 11$; therefore,

$$1\ 322\ 314\ 049\ 613\ 223\ 140\ 496 = \left[\dfrac{10^{11}+1}{11} \times 4\right]^2 =$$

$= 363\ 636\ 364^2$ is the smallest perfect square, which can be written down in the decimal system by writing down two identical numbers side by side.

(12) If $\alpha \equiv \beta \pmod{m}$, then $\alpha^s \equiv \beta^s \pmod{m}$, therefore also $a_s\alpha^s \equiv a_s\beta^s \pmod{m}$. In addition it is obvious, that $a_0 \equiv a_0 \pmod{m}$.

Adding the last congruence to the congruences $a_s\alpha^s \equiv a_s\beta^s \pmod{m}$ (for $s = 1, 2, \ldots n$) term by term we obtain $a_0 + a_1\alpha + a_2\alpha^2 + \ldots + a_n\alpha^n \equiv a_0 + a_1\beta + a_2\beta^2 + \ldots + a_n\beta^n \pmod{m}$ or $f(\alpha) \equiv f(\beta) \pmod{m}$.

(13) Since $8^2 \equiv -1 \pmod{5}$, therefore, by reasoning in the same way as in deducing the test for divisibility by 7 in the decimal system, we obtain the fact that the number M written down in the system of notation with base eight, is divisible by 5, if the algebraic sum of two-digit divisions of the number N is divisible by 5 (and conversely!). In order to obtain the remaining signs of divisibility it is sufficient to take into account,

that $8^2 \equiv -1 \pmod{13}$, $5^2 \equiv -1 \pmod{13}$; $5^2 \equiv 1$ $\pmod 8$; $3 \equiv 1 \pmod 2$; $3 \equiv -1 \pmod 4$; $3^3 \equiv -1$ $\pmod 7$.

(14) It is sufficient to find the smallest positive two-digit (or single-digit) number congruent, modulo 100, with the given number.

For example, $293^{293} \equiv (-7)^{293} \equiv -7 \times 49^{146} \equiv -7 \times 2401^{73} \equiv 93 \times 1^{73} \equiv 93 \pmod{100}$.

Therefore $293^{293} - 93$ is divisible by 100 (293^{293} gives the remainder 93, when divided by 100).

(15) Among the numbers 1, 2, 3, 4, ... $p^k - 2$, $p^k - 1$, p^k, the following are divisible by p: p, $2p$, $3p$, ... $p^{k-1}p = p^k$ (altogether p^{k-1} numbers). The remaining $p^k - p^{k-1}$ numbers are relatively prime to p.

(16) If $\varphi(n)$, on being divided by z_0 were to give the quotient q and the remainder r $(0 < r < z_0)$, i.e. $\varphi(n) = qz_0 + r$, then it would follow from $k\varphi(n) \equiv 1 \pmod n$, that

$$k^{qz_0+r} \equiv (k^{z_0})^q k^r \equiv 1^q \times k^r \equiv 1 \pmod n.$$

and this is impossible, since $r < z_0$.

(17) Suppose that the numbers m and n are written down in the base-k system of notation and $m < n$. If $k^{z_0} \equiv 1 \pmod n$ (z_0 is the smallest positive number possessing this property) then it also holds that $mk^{z_0} \equiv m \pmod n$. This means that by writing down, to the right of m, z_0 zeros (which is equivalent to multiplying m by k^{z_0}) and by dividing the number thus obtained by n, we have m as the remainder, and a certain group of digits. $c_1 c_2 \ldots c^{z_0}$ as quotient. A further adding on of z_0 zeros to the remainder again gives the same group of digits in the quotient, etc.

(18) (1) Substituting $x_0 + u$ in place of x, and $y_0 + v$ in place of y in the equation $ax + by = c$, we get $ax_0 + au + by_0 + bv = c$. Since $ax_0 + by_0 = c$, therefore $au + bv$ must be equal to zero, whence $au = -bv$; therefore au should be divisible by b. But a and b are relatively prime, therefore u must be divisible by b.

i.e. $u = bt$ (t may be any integer); then $v = -at$. And so $x_0 + bt$ and $y_0 \cdot - at$, t being any integer, satisfy the given equation.

(2) For an integral x and y, $ax + by$ is divisible by (a, b) and therefore cannot equal the number c, which is not divisible by (a, b), as was given.

(19) Since $\sqrt{2} = [1, 2, 2, 2, 2, \ldots]$ and $\sqrt{3} = [1, 1, 2, 1, 2, 1, 2 \ldots]$, the continued fractions for $= \sqrt{2}$ are

$$\frac{1}{1}, \frac{3}{2}, \frac{7}{5}, \frac{17}{12}, \frac{41}{29}, \frac{99}{70}, \frac{239}{169}, \frac{577}{408}, \frac{1393}{985}, \frac{3363}{2378}, \ldots,$$

and for $\sqrt{3}$

$$\frac{1}{1}, \frac{2}{1}, \frac{5}{3}, \frac{7}{4}, \frac{19}{11}, \frac{26}{15}, \frac{71}{41}, \frac{97}{56}, \frac{265}{153}, \frac{362}{209}, \frac{989}{571}, \frac{1351}{780}, \frac{3691}{2131}, \ldots$$

Therefore, $\sqrt{2} \simeq \dfrac{1393}{985} = 1 \times 4142131 \ldots$ and $\sqrt{3} \simeq$

$\simeq \dfrac{1351}{780} = 1 \cdot 7320512 \ldots$, here $\left| \sqrt{2} - \dfrac{1393}{985} \right| < \dfrac{1}{985 \times 2378}$

and $\left| \sqrt{3} - \dfrac{1351}{780} \right| < \dfrac{1}{780 \times 2131}$.

(20) From the first three equations with four unknowns, it is possible to express X, Y and Z by U and get $\dfrac{X}{1602} = \dfrac{Y}{891} = \dfrac{Z}{1580} = \dfrac{U}{2226} = s$, where s is any number.

From the last four equations it is possible to express x, y, z and u by X, Y, Z, U. If we then take $s = 4657t$, we obtain the values of X, Y, Z, U, x, y, z, u given in the text.

(21) If ABC is a "heronic triangle" (see Fig. 4) the lengths of the segments AD, BD, DC are rational, since $BD = \dfrac{2 \times S5ABC}{AC}$ $AD = \dfrac{AC^2 + AB^2 - BC^2}{2 \times AC}$ and $DC = = |AC - AD|$.

(22) The pythagorean triples (3, 4, 5) and (5, 12, 13) give rise to the heronic triples (25, 39, 56), (25, 39, 16) (25, 52, 63), (25, 52, 33), (20, 13, 21), (20, 13, 11), (15, 13, 14), (15, 13, 8).

The pythagorean triples (7, 24, 25) and (7, 24, 25) (give rise to three heronic triples only: (25, 25, 48), (25, 25, 14), (175, 600, 527), (the fourth "heronic triangle", made up of two triangles, similar to the given "pythagorean triangles", has sides 7, 24 and 25).

(23) b) $55\,555\,555 = 10\,001 \times 5555 = (7778+2223) \times$ $\times (7778-2223)$;   d) $12\,345\,678\,987\,654\,321 =$ $= 111\,111\,111^2$;

$1+2+3+4+5+6+7+8+9+8+7+6+5+4+3+$ $+2+1 = 81 = 9^2$.

(23a) $-\log_4 \left\{ \log_4 \underbrace{\sqrt{\sqrt{\sqrt{\ldots\sqrt{\sqrt{4}}}}}}_{} \right\} =$

$$= -\log_4 \{\log_4 4^{\frac{1}{2^{2n}}}\} = -\log_4 \left(\frac{1}{4^n}\right) = n.$$

(24) If the number x, $1 \leqslant x \leqslant 60$ is situated in columns numbered α, β, γ, in the tables shown in Fig. 4, then

$$x \equiv \alpha \ (\text{mod } 3), \tag{1}$$
$$x \equiv \beta \ (\text{mod } 4), \tag{2}$$
$$x \equiv \gamma \ (\text{mod } 5). \tag{3}$$

From (1) it follows that

$$x \equiv 3y + \alpha. \tag{4}$$

Substituting this expression in (2) we get $3y + \alpha \equiv \beta$ (mod 4), or $9y + 3\alpha \equiv 3\beta$ (mod 4), whence $y \equiv 3\beta - 3\alpha \equiv \alpha - \beta$ (mod 4) or

$$y = \alpha - \beta + 4z. \tag{5}$$

(On multiplying both sides of the congruence by a number mutually prime with the modulo, we obtain a congruence "equivalent" to the initial one — they both have the same roots: try to prove this.)

Substituting (5) in (4) we get

$$x = 3\,(\alpha-\beta)+12z+\alpha=4\alpha-3\beta+12z. \tag{6}$$

From (6) and (3) we have $4\alpha - 3\beta + 12z \equiv \gamma$ (mod 5) or $12\alpha - 9\beta + 36z \equiv 3\gamma$ (mod 5), whence $z \equiv 3\gamma - 12\alpha + 9\beta \equiv 3\gamma + 3\alpha + 4\beta$ (mod 5).

Therefore, $z = 3\gamma + 3\alpha + 4\beta + 5t$, which on substitution in (6) gives

$$x = 4\alpha - 3\beta + 12\,(3\gamma + 3\alpha + 4\beta + 5t) =$$
$$= 40\alpha + 45\beta + 36\gamma + 60t, \qquad (7)$$

or $x \equiv 40\alpha + 45\beta + 36\gamma$ (mod 60).

(25) Suppose it follows from $n \equiv d_1$ (mod 11) that $n^3 \equiv d_1^3 \equiv d$ (mod 11), where $0 \leqslant d_1 \leqslant 10$ and $0 \leqslant \leqslant d \leqslant 10$, i. e. d_1 is the remainder in dividing n by 11, and d is the remainder in dividing n^3 by 11.

It is easy to verify, that when d_1 equals 0, 1, 2, 3, 4, 5, 6, 7, 8, 9, 10, the corresponding values of d are 0, 1, 8, 5, 9, 4, 7, 2, 6, 3, 10, i. e. to each of the eleven values of d there corresponds "its own" value of d_1 different from others.

If we write out the values of d in order: 0, 1, 2, 3, 4, 5, 6, 7, 8, 9, 10, the corresponding values of d_1 are 0, 1, 7, 9, 5, 3, 8, 6, 2, 4, 10.

(26) For verification, it is sufficient to take into account, that: (1) the volume of water in all oceans is less than 1.4×10^{21} litres (7×10^{21} tumblersfull); (2) a tumblerfull contains $\dfrac{200}{18}$ gram-molecules of water, therefore the number of "marked" molecules is equal approximately to $\dfrac{6 \times 10^{23} \times 200}{18}$, or $\dfrac{2}{3}\,2 \times 10^{25}$.

(27) A light year is the distance L "covered" by light in one year

$L = 365 \times 25 \times 24 \times 60 \times 60 \times 3 \times 10^{10}$ cm $< 95 \times 10^{16}$ cm $< 10^{18}$ cm.

If v is the volume of a cube of edge $7 \times 10^7\,L$, and if N is the number of molecules of water, necessary to fill such a cube then

$v < (95 \times 10^{16} \times 7 \times 10^7)^3$ cm$^2 = (0.665 \times 10^{26})^3$ cm$^3 < \left(\dfrac{2}{3}\right)^2 \times$
$\times\ 10^{78}$ cm^3
and

$$N < \times\ \frac{6 \times 10^{23}}{8} \cdot \left(\frac{2}{3}\right)^3 \times\ 10^{78} = \frac{80}{81} \times 10^{100}.$$

(27a) Let $m = 4^{256}$: log $m = 256 \times \log 4 > 256 \times \times 0.602055 = 154.12608$ and $m > 1.336 \times 10^{154}$. Therefore, $Q = 4^m > 4^{1.336 \times 10^{154}}$ and log $Q > 1.336 \times \times 10^{154} \times 0.602055 > 0.80434 \times 10^{154} > 8 \times 10^{153}$ i.e. $Q > 10^{8 \times 10^{154}}$

(28) The radius of the sphere $R = \dfrac{10^{30}}{2} L < \dfrac{10^{30}}{2} \times 95 \times 10^{16}$ cm — see note([27]). The volume $v < \dfrac{4}{3}\pi \times 47.5^3 \times 10^{138}$ cm³ $< \dfrac{1}{2} \times 100^{144}$ cm³.

([29]) $\log_{1.000001} (e^{31 \times 10^6}) = 31 \times 10^6 \log_{1.000001} e \simeq 31 \times 10^6 \times \times 10^6 = 3 \times 1 \times 10^{13}$: $(e^{31 \times 10^6})^{10^{-6}} = e^{31} \simeq 100^{0.43 \times 31} = 10^{13.33} \simeq \simeq 2.2 \times 10^{13}$; $\log_{1.000001} (e^{32 \times 10^6}) \simeq 3.2 \times 10^{13}$; $(e^{32 \times 10^6})^{10^{-6}} = = e^{32} \simeq 5.86 \times 10^{13}$.

(30) Let, for example $n = 10\,000$. It follows from Stirling's formula, that
$$\log \sqrt{2\pi} \times 10^4 - 10\,000 \log e + 40\,000 < \log (10\,000!) <$$
$$< \log \sqrt{2\pi} \cdot 10^4 - 10\,000 \log e + 40\,000 + \frac{\log e}{120\,000}.$$
Since $\log e = 0.4342945$, $\log \pi = 0.49715$, $\log 2 = = 0.30103$ and $\dfrac{\log e}{120000}$ is a very small number, therefore $\log (10\,000!) = \dfrac{1}{2} (0.49715 + 0.30103 + 4) - 4342.945 + + 40\,000 = 35659.454$ i.e. $10\,000!$ is a 35 660-digit number.

(31) If we suppose that there exist two integers k and k', such that $a\dfrac{\sqrt{5}-1}{2} < k < (a=1)\dfrac{\sqrt{5}-1}{2}$ and $a\dfrac{3-\sqrt{5}}{2} < k < (a+1)\dfrac{3-\sqrt{5}}{2}$, then on adding these inequalities term by term we obtain $a < k + k' < a + 1$, which is impossible (a and $a+1$ are neighbouring integers). Obviously, a certain integer s is trapped between the numbers $a\dfrac{\sqrt{5}-1}{2}$ and $a\dfrac{\sqrt{5}-1}{2} + 1$. If s is smaller than $a\dfrac{\sqrt{5}-1}{2} + \dfrac{\sqrt{5}-1}{2}$, it is to be found in the first interval, if, on the other hand, $a\dfrac{\sqrt{5}-1}{2} + \dfrac{\sqrt{5}-1}{2} < s < a\dfrac{\sqrt{5}-1}{2} + 1$, then, on rewriting these inequalities in

the form $(a+1)\left(1 - \dfrac{3-\sqrt{5}}{2}\right) < s < ax\left(1 - \dfrac{3-\sqrt{5}}{2}\right) + 1,$

we obtain $(a+1)\dfrac{3-\sqrt{5}}{2} > a + 1 - s > a\dfrac{3-\sqrt{5}}{2}$, i. e. the second interval contains the integer $a + 1 - s$.

(32) Let us compile a table of special positions (c_k, d_k), using three rules, given at the beginning of § 11.

k	0	1	2	3	4	5	6	7	8	9	10	11	12	13	14	15	16	17	18	19	20	21	22	23	24	25	...
c_k	0	1	3	4	6	8	9	11	12	14	16	17	19	21	22	24	25	27	29	30	32	33	35	37	38	40	...
d_k	0	2	5	7	10	13	15	18	20	23	26	28	31	34	36	39	41	44	47	49	52	54	57	60	62	65	...

Correct moves: $(27, 37) \to (16, 26)$; $(14, 90) \to (14, 23)$; $(47, 69) \to (47, 29)$.

(33) It is easy to verify that for $a = 40, 55, 140$, the intervals $(a \times 0.618\ldots, (a+1)\ 0.618\ldots)$ contains the integers 25, 34, 87, and for $a = 400$, the integer 153 is contained in the interval $(a \times 0.381\ldots, (a+1)\ 0{\cdot}381\ldots)$. Therefore $40 = c_{25}$ $(d_{25} = c_{25} + 25 = 65)$, $55 = c_{34}$ $(d_{34} = c_{34} + 34 = 89)$, $140 = c_{87}$ $(d_{87} = c_{87} + 87 = 227)$, $400 = d_{153}$ $(c_{153} = 400 - 153 = 247)$.

(34) Correct moves: $(10, 17, 25) \to (8, 17, 25)$; $(47, 99, 181) \to (47, 99, 76)$; $(25, 43, 50)$ is a special position and all moves lead to defeat; from the position $(29, 29, 18)$ it is possible to move to $(15, 29, 18)$, or to $(29, 29, 0)$, and from $(93, 29, 74)$ to one of the positions $(87, 29, 74)$, $(93, 23, 74)$, $(93, 29, 64)$.

(35) As an example, we shall examine the situation, in which the rings numbered 12, 9, 7, 6, 2 are off, and the remaining rings are on the pin.

In order to take off 11, we must take off 8, 5, 4, 3, 1, after which it remains to make $1 + u_9 + u_{10}$ moves (take off 11, take up 1 to 9, take off 1 to 10).

But in order to take off 8, we must pass through the situation, when 1 to 7 are raised, as one of the intermediate stages, and for this we must take off, 1, 3, 4 (u moves), raise 6 (one move), take off 5 ($u_4 + u_5$

moves), raise 7 (one move) raise 1 to 5 (u_5 moves):
for all this we use $2u_5 + u_4 + 9$ moves altogether.
Adding to it u_8 (to take off 1 to 8) and the $1 + u_9 + u_{10}$
mentioned above we get altogether

$$u_{10} + u_9 + u_8 + 2u_5 + u_4 + 10 \text{ moves}$$

(36) To verify that the formula $u_n = 2^n - 1$ is valid,
it is sufficient (1) to show that it holds for $n = 1$,
(2) assuming that the formula holds for $n = k$ (k is
any natural number), to prove that it also holds for
$n = k + 1$.

The first follows from the fact that one lamina can
be transferred from the column A to the column B
in one move, i.e. $u_1 = 1 = 2^1 - 1$. Supposing now,
that $u_k = 2^k - 1$ we use the formula, proved in § 12,
$u_n = 2u_{n-1} + 1$, and we get $u_{k+1} = 2u_k + 1 = 2(2^k - 1) + 1 = 2^{k+1} - 1$.

(37)

a) 55—57, 75—55, 54—56, 74—54, 53—55, 73—53, 43—63,
51—53, 63—43, 33—53, 41—43, 53—33, 23—43, 31—33,
43—23, 13—33, 15—13, 25—23, 34—32, 13—33, 32—34,
45—25, 37—35, 57—37, 34—36, 37—35, 25—45, 56—54,
54—34, 46—44, 44—24.

b) 53—55, 73—53, 75—73, 65—63, 52—54, 73—53, 54—52,
51—53, 31—51, 32—52, 43—63, 51—53, 63—43, 45—65,
57—55, 65—45, 35—55, 47—45, 55—35, 25—45, 37—35,
45—25, 15—35, 13—15, 23—25, 34—36, 15—35, 36—34,
33—53, 34—54, 54—52.

(38) If we move the pieces along the line shown in
Fig. 130a which forms a closed "ring", one of whose
squares is "empty", the arrangement of the pieces with
respect to each other *along this ring* does not alter.
We must arrange the pieces in the ring — if the starting
position is solvable — in the following order (see Fig.
15 III).

$$1, 2, 3, 4, 8, 12, *, 15, 14, 13, 9, 10, 11, 7, 6, 5 \quad (1)$$

(counting from the upper left square and proceeding
clockwise. * is the empty square). We shall say, that
here No. 2 follows No. 1, No. 3 follows No. 2, No. 15
follows No. 12, No. 1 follows No. 5, etc.

The mutual arrangement of pieces in the ring changes only when B is an empty square, and the piece from A is moved there (or vice versa): the piece which is being moved moves two places "leftwards" (or "rightwards") along the ring. For example, in order to shift No. 9

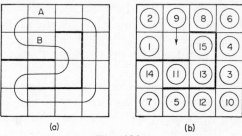

(a)   (b)

Fig. 130.

two places "leftwards", when the initial permutation is
$$5, 7, 14, 11, 13, 15, 1, 2, 9, 8, 6, 4, 3, 10, *, 12 \quad (2)$$
it is necessary to create, by means of a circular displacement of pieces, the situation b in Fig. 130, and then displace No. 9 into the empty square. This gives us
2, *, 8, 6, 4, 3, 10, 12, 5, 7, 14, 11, 13, 15, 9, 1.

Having then shifted No. 15 two places "leftwards" in an analogous manner, we manage to get No. 9 to follow No. 13, just like in the permutation (1).

Subsequently, we get, in the same way, No. 10 to follow No. 9, No. 11 to follow No. 10, No. 7 to follow No. 11, etc. Here if the permutation (2) is "solvable", we arrive finally at permutation (1) and if (2) is "nonsolvable" we arrive at the permutation 1, 2, 3, 4, 8, 12, *, 14, 15, 13, 9, 10, 11, 7, 6, 5, equivalent to IV in Fig. 15.

(39) For $n = 1$ and for $n = 2$ formula (6) in § 15 gives:

$$v_1 = 1 \frac{1}{\sqrt{5}} \left\{ \left(\frac{1+\sqrt{5}}{2} \right)^2 - \left(\frac{1-\sqrt{5}}{2} \right)^2 \right\} = 1, \qquad v_2 =$$

$$= \frac{1}{\sqrt{5}} \left\{ \left(\frac{1+\sqrt{5}}{2} \right)^3 - \left(\frac{1-\sqrt{5}}{2} \right)^3 \right\} = \frac{2\{3\sqrt{5}+(\sqrt{5})^3\}}{8\sqrt{5}} = 2.$$

Then

$$v_{s-1} + v_{s-2} = \frac{1}{\sqrt{5}}\left\{\left(\frac{1+\sqrt{5}}{2}\right)^s - \left(\frac{1-\sqrt{5}}{2}\right)^s\right\} +$$

$$+ \frac{1}{\sqrt{5}}\left\{\left(\frac{1+\sqrt{5}}{2}\right)^{s-1} - \left(\frac{1-\sqrt{5}}{2}\right)^{s-1}\right\} =$$

$$= \frac{1}{\sqrt{5}}\left\{\left(\frac{1+\sqrt{5}}{2}\right)^{s-1}\left(\frac{1+\sqrt{5}}{2} + 1\right)\right\} -$$

$$- \left(\frac{1-\sqrt{5}}{2}\right)^{s-1}\left(\frac{1-\sqrt{5}}{2} + 1\right)\right\} =$$

$$= \frac{1}{\sqrt{5}}\left\{\left(\frac{1+\sqrt{5}}{2}\right)^{s+1} - \left(\frac{1-\sqrt{5}}{2}\right)^{s+1}\right\} = v_s.$$

since

$$\frac{1+\sqrt{5}}{2} + 1 = \left(\frac{1+\sqrt{5}}{2}\right)^2 \text{ and } \frac{1-\sqrt{5}}{2} + 1 =$$

$$= \left(\frac{1-\sqrt{5}}{2}\right)^2.$$

(40) Using the three relationships connecting w_s, w_{s-1}, w_{s-2}, w_{s-3} in turn, and taking into account that $w_1 = w_2 = 1$ and $w_2 = 2$, it is easy to get

s	0	1	2	3	4	5	6	7	8	9	10	11	12	13	14	15	...
w_3		1	1	2	4	7	13	17	30	60	107	197	257	454	908	1619	...

by filling in the lower row of the table (the numbers printed in bold type are those of "the squares with sticky soil").

(41) Any particular method of transition from the node O $(0, 0, \ldots, 0)$ to the node A $(a_1, a_2, \ldots, a_m)$ where $a_1 + a_2 + \ldots + a_m = n$ can be characterized by an arrangement of n letters, among which the letter x_1 (indicating that the corresponding move increases the first coordinate of the point by one) is encountered a_1 times, the letter x_2 is encountered a_2 times, etc. and finally, the letter x_m is encountered a_m times. But a_1 places out of n can be filled by the letter x_1 in $C_n^{a_1}$ ways:

to each of these ways there correspond the remaining $n - a_1$ empty places. It follows that we can pick a_1 places for x_1 and a_2 places for x_2 (out of n places) in $C_n^{a_1}$, $C_{n-a_1}^{a_2}$ ways. To each of these ways there correspond $C_{n-a_1-a_2}^{a_3}$ ways of distributing a_3 letters x_3 in the remaining $n - a_1 - a_2$ places, i.e. there are altogether $C_n^{a_1} \times \times C_{n-a_1}^{a_2} \times C_{n-a_1-a_2}^{a_3}$ methods of placing letters x_1, x_2 and x_3.

Continuing this reasoning, we arrive at the result, that for the distribution a_1 letters x_1, a_2 letters x_2, ..., a_{m-1} letters x_{m-1} (letters x_m take up automatically the remaining empty a_m places), there are

$$C_n^{a_1} \cdot C_{n-a_1}^{a_2} \cdot C_{n-a_1-a_2}^{a_3} \; \cdots \; C_{n-a_1-a_2-\ldots-a_{m-2}}^{a_{m-1}} =$$
$$= \frac{n!}{a_1! \, (n-a_1)!} \cdot \frac{(n-a_1)!}{a_2! \, (n-a_1-a_2)!} \cdots$$
$$\cdots \frac{(n-a_1-a_2-\ldots-a_{m-2})!}{a_{m-1}! \, a_m!} = \frac{n!}{a_1! \, a_2! \ldots a_m!}$$

methods.

12	(2)	4	8	8	(8)	8	16
(10)	2	(2)	4	(8)	16	48	8
8	2	4	(2)	8	48	16	(8)
6	2	2	2	(4)	8	(8)	8
4	(2)	2	(8)	2	(2)	4	8
(2)	2	(4)	2	2	4	(2)	4
1	2	2	(2)	2	2	2	(2)
A	1	(2)	4	6	8	(10)	12

Fig. 131.

(42) Figure 131 shows in how many ways a rook can reach a particular square of a chessboard with "obstacles", in the least number of "short moves".

For convenience, the data referring to the squares of the fourth zone (2, 2, 4, 2, 2) of the eighth zone (10, 2, 2, 2, 4, 8, 2, 2, 2, 10) and the twelfth zone (8, 8, 8, 8) are encircled.

(43) Figure 132 shows, that the fourth zone consists of thirty-two squares. The king can get into the fourth zone in 320 ways (the sum of numbers in the squares of the fourth zone).

1	4	10	16	19	16	10	4	1
4	1	3	6	7	6	3	1	4
10	3	1	2	3	2	1	3	10
16	6	2	1	1	1	2	6	16
19	7	3	1		1	3	7	19
16	6	2	1	1	1	2	6	16
10	3	1	2	3	2	1	3	10
4	1	3	6	7	6	3	1	4
1	4	10	16	19	16	10	4	1

Fig. 132.

(44) Two pawns can be transferred from the second line to the eighth line in:

(*a*) 12 moves, without taking advantage of the right of the double move, in $\frac{12!}{6!\,6!}$ ways;

(*b*) 11 moves, making use of the right of the double move in the case of the first pawn only (or the second pawn only) in $\frac{11!}{5!\,6!}$ ways;

(*c*) 10 moves, making use of the right of the double move for both pawns, in $\frac{10!}{5!\,5!}$ ways. We have, for the two pawns, altogether $\frac{12}{6!\,6!} + 2\frac{11}{5!\,6!} + \frac{10}{5!\,5!}$ ways.

Similarly, we have, for three and four pawns,

$\frac{18!}{6!\,6!\,6!} + 3\frac{17!}{5!\,6!\,6:} + 3\frac{16!}{5!\,5!\,6!} + \frac{15!}{5!\,5!\,5!}$ ways and

$\frac{24!}{6!\,6!\,6!\,6!} + 4\frac{23!}{5!\,6!\,6:6!} + 6\frac{22!}{5!\,5!\,6!\,6!} + 4\frac{21!}{5!\,5!\,5!\,6!} +$

$+ \frac{20!}{5!\,5!\,5!\,5!}$ ways respectively.

(45) (*a*) A triangle cannot be constructed as it would need to have [see Fig. 29*e*] six directions, in each of which the sum of three numbers should be the same. But the number 12 can be represented in five ways only as a sum of three different numbers not exceeding seven; $12 = 1 + 4 + 7 = 1 + 5 + 6 = 2 + 3 + 7 =$

$= 2 + 4 + 6 = 3 + 4 + 5$ (there are only four ways of representing each of the numbers 11 and 13 as a sum of three numbers, there are four ways for 10 and 14, three ways for 9 and 15, etc.).

(*b*) It is not possible to construct a pentagon, since the number at its centre must be a part of five equal sums, each of which consists of three different numbers not exceeding eleven: but it is easy to verify directly that it is impossible to find such a number.

(46) (1) A generalized domino set has $n + 1$ pieces of the form (k, k) and $C_{n+1}^2 = \dfrac{(n+1)n}{2}$ pieces of form (l, m) where $l \neq m$, and the total number of pieces is $n + 1 + \dfrac{(n+1)n}{2} = \dfrac{(n+2)(n+1)}{2}$.

(2) Each number k is encountered n times in combination with numbers m not equal to k, and twice in the piece (k, k). Therefore, the sum of all points equals

$$(0 + 1 + 2 + \ldots + n)(n + 2) = \frac{n(n+1)(n+2)}{2}.$$

(47) If the piece (a, b) $(a \pm b)$ is removed from the complete set of generalized dominoes, when n is even, a points (as also b points) are encountered in the remaining pieces an odd number $(n + 1)$ of times (see([46]), 2).

On the other hand, evidently, any closed chain contains each number of points an even number of times; any number of points may be contained an odd number of times only in an open chain, and this number of points will then be at one of the ends of the chain.

(48) When n is odd each number of points is encountered an odd number $(n + 2)$ of times in a generalized domino set. In an open chain, only those numbers of points that are to be found at the ends of the chain are contained an odd number of times in the chain. Therefore, at least $n - 1$ of the numbers $0, 1, 2, 3, \ldots,$ $n - 1, n$ are to be found on the pieces, not included in the chain, i. e. at least $\frac{1}{2}(n - 1)$ pieces of the domino

set cannot be arranged in a chain. But there are alto-gether $\dfrac{(n+1)\,(n+2)}{2}$ pieces. Therefore a chain can be made up of at, most, $\dfrac{(n+1)\,(n+2)}{2} - \dfrac{n-1}{2} = \dfrac{n^2 + 2n + 3}{2}$ pieces.

(49) One of the possible distributions of pieces is as follows; Ist player; (0, 0), (0, 1), (0, 2), (0, 3), (1, 4), (1, 5), (1, 6); IVth player: (1, 1), (1, 2), (1, 3) ,(0, 4), (0, 5), (0, 6), (2, 2).

Players I and IV make a chain

(0, 0) (0, 4) (4, 1) (1, 2) (2, 0) (0,5) (5, 1) (1, 1) (1,0)
(0, 6) (6, 1) (1, 3) (3, 0)

(players II and III cannot enter the game, as they do not possess pieces containing zero points, or one point).

(50) Let us number the squares of a "4,5-board" (Fig. 39*b*) as shown in Fig. 133.

The following sequence of moves leads to the goal:

I	2	3	4
5	6	7	8
9	10	II	12
13	14	15	16
17	18	19	20

Fig. 133.

1.	19—14	2— 7	10.	17—11	4—10
2.	18—15	3— 6	11.	2—12	19— 9
3.	14— 8	7—13	12.	11—16	10— 5
4.	15—12	6— 9	13.	12— 7	9—14
5.	20— 5	1—16	14.	18—13	3— 8
6.	5— 2	16—19	15.	16— 6	5—15
7.	8—11	13—10	16.	7— 2	14—19
8.	12—18	9— 3	17.	13— 4	8—17
9.	11— 1	10—20	18.	6— 3	15—18

(51) Figure 119a, shows a plane arrangement, whose coverage according to Hamilton's rules is equivalent to the coverage of a "3, 4-board" by the chess knight. It is easy to verify that one cannot visit all points of the arrangement, moving along the segments and visit-ing each point once only, if one starts at any of the points *L, M, R, B, C, Q*. The possible circuits are

NBPMAQCKRDLS,   *PMAQNBRKCSLD,*
NBPMAQCSLDRK,   *PMAQNBRDLSCK,*
NQAMPBRKCSLD,   *AMPBNQCKRDLS,*
NQAMPBRDLSCK,   *AMPBNQCSLDRK,*

each of which represents an open polygon which can, naturally, be traversed also in the reverse order (which we do not regard as a new circuit).

(51 a) If we write down the first forty natural numbers in their order, and if we begin discarding every third number, beginning from left to right (by underlining it and indicating in brackets in which turn it was discarded) we obtain

1	2	3	4	5	6	7	8	9	10	11	12	13	14	15	16	17	18	19	20
(37)	(14)	(1)	(23)	(29)	(2)	(15)	(33)	(3)	(24)	(16)	(4)	(39)	(30)	(5)	(17)	(25)	(6)	(36)	(18)

21	22	23	24	25	26	27	28	29	30	31	32	33	34	35	36	37	38	39	40
(7)	(34)	(26)	(8)	(19)	(31)	(9)	(40)	(20)	(10)	(27)	(38)	(11)	(21)	(32)	(12)	(28)	(22)	(13)	(35)

whence it can be seen that the element No. 13 is the one-but-last to be discarded, and No. 28 is the last one.

(2) In our case $n = 40$, $k = 3$, $q = \dfrac{k}{k-1} = 3$, $nk = 120$. For $s = 39$ $a_1 = k(n-s)+1 = 4$ and the integral geometric progression consists of numbers 4, 6, 9, 14, 21, 32, 48, 72, 108, 162 .... . Since $162 > nk = 120$, therefore $A = 108$ and $t = nk + 1 - A = 120 + 1 - 108 = 13$, i. e. the thirteenth element is discarded 39th.

If $s = 40$, $a_1 = 1$ and the "integral geometric progression" consists of numbers 1, 2, 3, 5, 8, 12, 18, 27, 41, 62, 93, 140 .... .

Since $140 > nk = 120$, we have $A = 93$ and $t = 120 + 1 - 93 = 28$, i. e. the 28th element was discarded 40th.

(51 b)

1	2	3	4	5	6	7	8	9	10	11	12	13	14	15	16	17	18
(28)	(19)	(22)	(36)	(16)	(1)	(7)	(12)	(35)	(31)	(25)	(2)	(20)	(8)	(27)	(13)	(17)	(3)

19	20	21	22	23	24	25	26	27	28	29	30	31	32	33	34	35	36
(23)	(30)	(9)	(32)	(29)	(4)	(14)	(21)	(18)	(10)	(34)	(5)	(26)	(24)	(15)	(33)	(11)	(6)

It can be seen from the above table that the cards should be arranged in the following order: the ace of spades, (the twenty-eighth card or, in other words the first card of the fourth suit) the ace of clubs (card No. 19 — the first card of the third suit) the knave of clubs (card No. 22 — the fourth card of the third suit) etc.

Since $n = 36$, $k = 6$, $q = \dfrac{k}{k-1} = \dfrac{6}{5}$, $nk = 216$,

we obtain the following integral geometric progression

when $s = 19$: $a_1 = 6 \times 17 + 1 = 103$, 124, 149, 179, 215, 258 . . .

when $s = 31$: $a_1 = 6 \times 5 + 1 = 31$, 38, 46, 56, 68, 82, 99, 119, 143
172, 207, 249, . . .

when $s = 17$: $a_1 = 6 \times 19 + 1 = 115$, 138, 166, 200, 240, . . .

Since in these "progressions" the greatest numbers not exceeding the number nk are 215, 207, 200, the corresponding values of t are $2(216 - 215 + 1)$, $10(216 - 207 + 1)$ and $17(216 - 200 + 1)$ see the bold type figures in the table.

(52) If we distribute the elements a_1, a_2, . . ., a_k round a circle in a clockwise direction, and we then transfer each of them anti-clockwise into the place occupied by its neighbour, then this operation is equivalent to the cyclic permutation $C = (a_1, a_2, \ldots, a_k)$.

Evidently, $C^k = E$, since after k repetitions of this operation all elements find themselves once again in their original places.

If the permutation A is equal to the product of several independent cycles: $A = C_1 \cdot C_2 \ldots C_s$, whose orders are k_1, k_2, . . ., k_s, then, in the permutation A^m all elements will find themselves in their former places only in the case when m is divisible by k_1, by k_2, by k_3, etc. The smallest value m satisfying these conditions equals the smallest common factor of the numbers k_1, k_2, . . ., k_s.

(53) Suppose that in accordance with the rule of transition from one permutation to another, then to the next, etc., indicated in the text of § 22, in each transition the elements occupying, say, places numbered α_1, α_2, . . ., α_s are replaced by each other in a cyclic order. Then the permutation M will contain the cycle $(a'_1, a'_2, \ldots, a'_s)$, where a'_1, a'_2, . . ., a'_s are elements which had been occupying places numbered α_1, α_2, . . ., α_s to begin with.

Other cycles forming a part of the required permutation M are brought to light in a similar way.

(54) We denote the angles of the rhombi of the kth layer nearest to the centre by α_k. If the point A is common to the rhombi of the $(k-1)$th and the $(k+1)$th layers, then $\alpha_{k-1} + 2(\pi - \alpha_k) + \alpha_{+k1} = 2\pi$ (see Fig. 134a) whence $\alpha_{k+1} - \alpha_k = \alpha_k - \alpha_{k-1}$, i. e. $\alpha_1, \alpha_2, \alpha_3, \ldots$ form an arithmetical progression, whose common difference equals $\alpha_1 = \dfrac{2\pi}{m}$, since $\alpha_2 = 2\alpha_1$ (Fig. 134b), whence $\alpha_2 - \alpha_1 = \alpha_1 = \dfrac{2\pi}{m}$; therefore $\alpha_k = k\,\dfrac{2\pi}{m}$

If m is odd, then $\alpha\,\dfrac{^{1}/_{2}m - ^{1}/_{2}}{2} = \dfrac{m-1}{2} \times \dfrac{2\pi}{m} = \dfrac{\pi\,(2m-2)}{2m}$, i. e. $\alpha\,\dfrac{^{1}/_{2}m - ^{1}/_{2}}{2}$ equals the angle of a regular $2m$-sided polygon. Other angles of the $2m$-sided polygon are made up of three angles, whose sum is equal to

$$\alpha_{\frac{m-1}{2} - 1} + 2(\pi - \alpha_{\frac{m-1}{2}}) = \frac{2\pi}{m} \times \frac{m-3}{2} + 2\left(\pi - \frac{m-1}{2} \times \right.$$
$$\left. \times \frac{2\pi}{m}\right) = \frac{\pi}{m}\,(m-3+2) = \frac{\pi(m-1)}{m} = \alpha_{\frac{m-1}{2}}.$$

When m is even $\alpha\,\dfrac{m}{2} - 1 = \dfrac{2\pi}{m}\left(\dfrac{m}{2} - 1\right) = \dfrac{\pi\,(m-2)}{m}$

i. e. $\alpha\,\dfrac{m}{2} - 1$ equals the angle of a regular m-sided polygon: at the same time "the composite angle" (like the angle D in Fig. 54b) equals

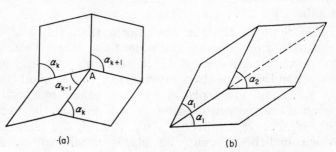

Fig. 134.

$$\alpha_{\frac{m}{2}-2} + 2\left(\pi - \alpha_{\frac{m}{2}-1}\right) =$$

$$= 2\pi + \left[\frac{m}{2} - 2 - 2\left(\frac{m}{2} - 1\right)\right] \times \frac{2\pi}{m} = \pi.$$

(55) If we denote the angle of the rhombus of the kth layer nearest to the centre of the regular polygon, by α_k, then it is easy to prove, that $\alpha_k = (2k+1) \cdot \frac{\pi}{m}$ (it is sufficient to establish, that $\alpha_1 = \frac{3\pi}{m}$ and $\alpha_2 = \frac{5\pi}{m}$ and to make use of the self-evident equality $\alpha_{k-1} + \alpha_{k+1} + 2(n - \alpha_k) = 2\pi$, whence it follows that α_1, α_2, α_3, α_4, etc., form an arithmetic progression). But then $\alpha_{\frac{1}{2}(m-3)} = \left\{2\left(\frac{m-3}{2}\right) + 1\right\} \frac{\pi}{m} = \frac{\pi(m-2)}{m}$ and this is the angle of a regular m-sided polygon. In addition, "the composite angle" (like the angle with vertex E in Fig. 55b) equals

$$\alpha_{\frac{m-5}{2}} + 2(\pi - \alpha_{\frac{m-3}{2}}) = (m-4)\frac{\pi}{m} + 2\left[\pi - \right.$$

$$\left. - \frac{\pi(m-2)}{m}\right] = \pi.$$

It is easy to verify that each of the angles of the "open polygon" (like the polygon $ABCDEFGHIKL$ in Fig. 55b equals $\frac{\pi(m-2)}{m}$, whence it follows that the "open polygon" is regular (for example, if α is the angle at the vertex of a starlike polygon, then $\triangle\, ABC = \alpha + (\pi - \alpha_1) = \alpha + \pi - 3\alpha = \pi - 2\alpha$, $\triangle\, BCD = \alpha_1 + (\pi - \alpha_2) = 3\alpha + \pi - 5\alpha = \pi - 2\alpha$, etc., and $\pi - 2\alpha = \pi - \frac{2\pi}{m} = \frac{\pi(m-2)}{m}$, which is the angle of a regular m-sided polygon.

(56) If we take the sides of the constituent squares to equal 2 cm, 5 cm, 7 cm, etc., the areas of the rectangles S_1, S_2, S_3, composed of nine, ten and thirteen squares respectively, are

4209 cm² ($2^2+5^2+7^2+9^2+16^2+25^2+28^2+33^2+36^2$),
10270 cm² ($3^2+11^2+12^2+23^2+34^2+35^2+38^2+$
$+41^2+44^2+45^2$),
27495 cm² ($1^2+4^2+5^2+9^2+14^2+19^2+33^2+52^2+$
$+56^2+69^2+70^2+71^2+72^2$).

Since $4209 = 3 \times 23 \times 61$, the sides of the rectangle S, can equal either 3 cm and 1403 cm, or 69 cm and 61 cm, or 23 cm and 183 cm. The first and the third cases do not apply, as no side of S_1 must be less than the side of the largest of the constituent squares (36 cm). Therefore the sides of the rectangle S_1 are 69 cm and 61 cm.

Similarly $10\,270 = 2 \times 5 \times 13 \times 79 = 79 \times 130 = 65 \times 158$. But 65 cm cannot be represented as a sum of sides of squares of given dimensions. Therefore the sides of S_2 are 79 cm and 130 cm. The sides of the rectangle S_3 are 141 cm and 195 cm, for $27\,495 = 3 \times 3 \times 5 \times 13 \times 47 = 141 \times 195$ (other ways of factorizing the number 27 495 to give two factors do not agree with the dimensions of the constituent squares).

If we write "square a" for "square of side a", then S_1 consists of (reading from left to right) squares 36, 33 (upper row) 5, 28 (underneath 33); 25, 9, 2 (underneath 36); 7 (underneath 2 and 5), 16 (underneath 9 and 7); S_2 consists of squares 41, 44, 45 (upper row); 38, 3 (underneath 41); 35, 12 (underneath 3 and 44); 11, 34 (underneath 45); 23 (underneath 12 and 11); S_3 consists of squares 71, 72, 52 (upper row): 19, 33 (underneath 52); 5, 14 (underneath 19); 70, 1 (underneath 71); 69, 4 (underneath 1 and 72); 9 (underneath 4 and 5); 56 (underneath 9, 14, 33).

(57) "Rectangle 608 × 377" consists of squares 209, 205, 194 (upper row), 11, 183 (underneath 194); 44, 172 (underneath 205 and 11); 168, 41 (underneath 209); 1, 43 (underneath 44;) 42 (underneath 41 and 1); 85 (underneath 42 and 43). "Rectangle 608 × 231" consists of squares 231, 95, 61, 108, 113 (upper row), 34,

27 (underneath 61); 7, 20 (underneath 27); 136 (underneath 95, 34 and 7); 123, 5 (underneath 128); 118 (underneath 5 and 113).

(58) If the smallest square $ABCD$ were to adjoin (along its side AB) one of the sides of a rectangle, then it would be held between two larger squares, or it would be "pressed" by a larger square to the neighbouring side of the rectangle; in both cases no square of larger dimensions could adjoin the side CD of the smallest square.

(59) (1) $a^3 = \left(\frac{2a}{3}\right)^3 + 17\left(\frac{a}{3}\right)^3 + 16\left(\frac{a}{6}\right)^3$, i. e. a cube of edge a can be split up into 34 cubes: one cube of edge $\left(\frac{2a}{3}\right)$ 17 cubes of edge $\left(\frac{a}{3}\right)$ and 16 cubes of edge $\left(\frac{a}{6}\right)$.

(2) $a^3 = a\left(\frac{a}{2}\right)^3 + 48\left(\frac{a}{4}\right)^3$ i. e. a cube of edge a can be split up into 50 cubes: 2 cubes of edge $\left(\frac{a}{2}\right)$ and 48 cubes of edge $\left(\frac{a}{4}\right)$.

(60) A square of side a can be split up into 4 squares, into 6 squares $\left[\text{one of side } \left(\frac{2a}{3}\right), \text{ five of side } \left(\frac{a}{3}\right)\right]$ into 8 squares $\left[\text{one of side } \left(\frac{3a}{4}\right), \text{ seven of side } \left(\frac{a}{4}\right)\right]$.

But if a square can be subdivided into s parts, then, by subdividing any one of the squares obtained into four parts we can subdivide the original square into $s + 3$ parts. Therefore, a square can be subdivided into seven $(4 + 3)$ into ten $(7 + 3)$, etc., and, in general, into $4 + 3k$ squares. By similar reasoning, we show that a large square can be subdivided into $6 + 3l$ squares and $8 + 3m$ squares (k, l, m are natural numbers).

But any natural number n, beginning with 6 can be represented in one of the forms $6 + 3l$, $4 + 3k$, $8 + 3m$, since all these numbers, when divided by three, leave a remainder, which is either 0·1 or 2.

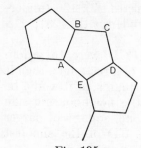

Fig. 135.

(61) All vertices of some regular decagon should be nodes of the type (10, 5, 5) — see Fig. 135. But then points B and D cannot be nodes of the same type, since in this case two regular decagons and one regular pentagon would "meet" at C, which is impossible.

(62) Each angle of a regular m-sided polygon equals $\frac{190° \, (m-2)}{m}$. The sum of angles, whose vertices are at the node $(n_1, n_2, \ldots, n_k)$ equals 360°, therefore

$$\frac{180° \, (n_1-2)}{n_1} + \frac{180° \, (n_2-2)}{n_2} + \ldots + \frac{180° \, (n_k-2)}{n_k} = 360°.$$

Dividing throughout by 180° we get $k - 2 \left(\frac{1}{n_1} + \frac{1}{n_2} \ldots + \frac{1}{n_2} \right) = 2.$

(63) Figure 136a shows that it is possible to construct a hexagon $KCLMNP$, whose opposite sides are parallel and equal, out of four pentagons of form $ABCDE$, $(AB = BC = CD = DE; \, <B = <D = 90°; \, <A = = <C = <E = 120°)$ and it is easy to cover a plane with such hexagons. Figure 136b shows that a plane can be covered with regular pentagons and rhombi of acute angle 36°.

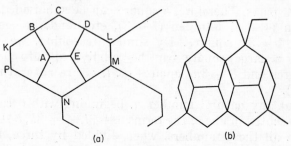

(a)   (b)

Fig. 136.

(64) It is sufficient to prove that whether we begin by reflecting the triangle ABC in side BC or in the side AB, the arbitrary point M in the triangle ABC produces, finally, the same set of five points M_1, M_2, M_3, M_4, M_5 (Fig. 137).

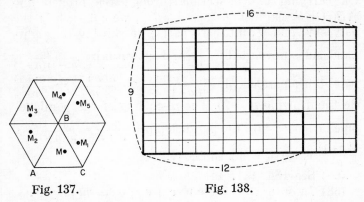

Fig. 137.   Fig. 138.

(65a) See Fig. 138.

(65b) It is sufficient to draw straight lines, parallel to the sides of the rectangle and dividing side a into $n + 1$ parts and side b into n parts, and to draw the broken line as shown in Fig. 139 (in this drawing $n = 6$).

(66) Figure 140 shows a parallelepiped $\dfrac{am}{m+1} \times$

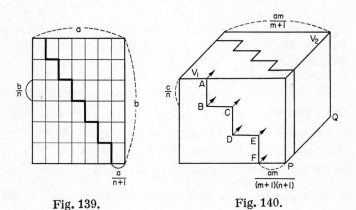

Fig. 139.   Fig. 140.

$\times \dfrac{b\,(m+1)}{m} \times c$" composed of two halves, V_1 and V_2 of the parallelepiped "$a \times b \times c$" (in the drawing $m = 4$ and $PQ = \dfrac{b\,(m+1)}{m}$.

On carrying out a section through the broken line $ABCDEF$, parallel to the edge PQ, we cut each of the halves V_1 and V_2 into two parts; on shifting the right-hand sides upwards by $\left(\dfrac{c}{n}\right)$ and leftwards by $\dfrac{am}{(m+1)\,(n+1)}$ we obtain a paralellepiped $\dfrac{amn}{(m+1)(n+1)} \times \dfrac{b\,(m+1)}{m} \times \dfrac{c\,(n+1)}{n}$ (in the drawing $n = 3$). The re-cutting of the paralellepiped $a \times b \times c$ into the parallelepiped $\dfrac{am}{m+1} \times \dfrac{b\,(m+1)n}{m\,(n+1)} + \dfrac{c\,(n+1)}{n}$ is carried out similarly.

(67) See Fig. 141.

(68) In order that the curve $\begin{cases} x = a \sin mt \\ y = b \sin nt \end{cases}$ may pass through one of the vertices of a rectangle "circumscribed" about the curve, it is required to have, for some value of the parameter t, $mt = (2k + 1) \times 90°$ and $nt = (2l + 1) \times \times 90°$, where k and l are some integers, and that is possible only on condition that integers k and l may be found, for which $\dfrac{m}{n} = \dfrac{2k+1}{2l+1}$.

In this case, for $t = \dfrac{(2k+1)90°}{m} = \dfrac{(2l+1)90°}{n} = t_0$ we have $|x_0| = |y_0| = 1$. We leave it to the reader to verify that points $(x_1,\ y_1)$ and $(x_2,\ y_2)$ corresponding to the values of the parameter $t_1 = t_0 - \varDelta t$ and $t_2 = t_0 + \varDelta t$ ($\varDelta t$ is arbitrary) coincide.

(69) I. If we do not utilize the sign of absolute value, the equation $|(2y - 1)| + |2y + 1| + \dfrac{4\,|x|}{\sqrt 3} = 4$ is written down differently in each of the regions A_1, A_2, A_3, A_4, A_5, A_6 (Fig. 142a). For example, when $y \geqslant \frac{1}{2}$ and $x \leqslant 0$ (region A_2) we have $2y - 1 + 2y + 1 + \dfrac{4(-x)}{\sqrt 3} = = 4$, or $y - \dfrac{x}{\sqrt 3} = 1$. This equation corresponds to the

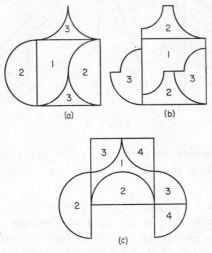

Fig. 141.

straight line l, from which the segment KL, lying in the region A_2, is taken. When $y \leq \frac{1}{2}$ and $x \geq 0$, we have $1 - 2y + 2y + 1 + \frac{4x}{\sqrt{3}} = 4$, or $x = \frac{\sqrt{3}}{2}$, and this is the equation of the straight line m, from which only the segment MN lying in the region A_6, is taken etc.

II. If the sign of absolute value is not used, the equation $|x| + |y| + \frac{1}{\sqrt{2}} \left\{ |x - y| + |x + y| \right\} = 2 + 1$ is written down differently in each of the eight "sectors" B_k $(1 \leq k \geq 8)$, into which a plane is divided by the coordinate axes and by the bisectors of angles between the coordinates (see Fig. 142b). For example, in sector, B_4, where $x \leq 0$, $y \geq 0$, $x - y \leq 0$, $x + y \leq 0$, we have $-x + y + \frac{1}{\sqrt{2}} \left\{ y - x + (-x - y) \right\} = 1 + \sqrt{2}$, or $y - (1 + \sqrt{2})x = 1 + \sqrt{2}$, and this is the equation of the straight line l, passing through the point $(-1, 0)$ at an angle $67°30'$ to the axis Ox ($\tan 67°30' = 1 + \sqrt{2}$). The part PQ of the straight line, which lies in the sector B_4 gives the side BQ of a regular octagon. We leave

273

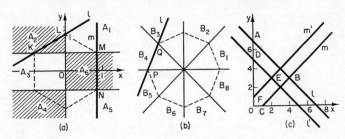

Fig. 142.

it to the reader to prove that the sides of a regular octagon lying in other sectors are obtained in the same way.

III. Since x and y enter the equation $|x| + ||y| - 3| - 3| = 1$ only within the sign of absolute value, it is sufficient to construct a part of the "curve", lying in the first quarter, then to construct "curves" symmetrical to it with respect to the axis Ox, the axis Oy and the origin of the coordinates.

For $x \geq 0$ and $y \geq 0$, we have $|x| + |y-3|-3 = 1$, whence it either follows, that $x + |y-3|-3 = 1$.

$$|y-3| = 4-x, \tag{1}$$

or $x + |y-3| -3 = -1$, this is

$$|y-3| = 2-x. \tag{2}$$

The eqn. (1) makes sense only when $x \leq 4$, and from it follows either $y-3 = 4-x$ (the straight line l, from which the segment AB, where $x \leq 4$, should be taken) or $y - 3 = x - 4$ (the straight line m, from which the segment BC has to be taken, for $x \leq 4$) (see Fig. 142c).

The eqn. (2) gives $y - 3 = 2 - x$ (the straight line l' from which the segment DE, where $x \leq 2$, must be taken) or $y - 3 = x - 2$ (the straight line m', from which the segment FE, where $x \leq 2$, should be taken).

(70) Let $y = f(x)$ be the equation of the curve l and let the curve l' be shifted to the right by α with

respect to l, (Fig. 143). If $a'(x, y)$ is any point on the curve l', and $A(X, Y)$ the corresponding point of the curve l, i. e. $X = x$ and $Y - y$, then $Y = f(X)$ and therefore

$$y = f(x - \alpha),$$

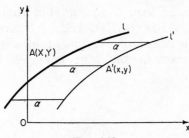

Fig. 143.

this is the equation of the curve l', which is satisfied by the coordinates of any point on it.

(70a) Seven-figure tables of logarithms should be used.

(71) (1) Six planes, passing through the edges of a regular tetrahedron and equally inclined to the corresponding faces of the tetrahedron, form at their intersection a cube. (see Fig. 100b).

(2) Twelve long diagonals of a rhombic dodecahedron (O_1, O_2, etc. — see Fig. 105) serve as the edges of a certain regular octahedron. Planes passing through the edges of this octahedron and equally inclined to its corresponding faces, bound a rhombic dodecahedron, depicted in Fig. 105.

(72) Suppose, that two colours are sufficient to colour regions into which the plane is divided by n straight lines. If, on drawing the $(n + 1)$th straight line, we preserve the colour of all regions (and portions of regions) of the "nth-subdivision" on one side of the $(n + 1)$th line, and on the other side we change the colour of each region into the other one, then each two bordering regions of the $(n + 1)$th subdivision are coloured differently. Since two colours are sufficient,

when $n = 1$, and we have proved the transition from n to $n + 1$, therefore it is possible to colour a plane sub-divided into regions by any number of straight lines, using two colours only.

These arguments hold also in the case of space broken up by n planes.

(73) It can be seen from Fig. 144 that any two of four triangles *ABD*, *BCE*, *CAF*, *DEF* have a common border in the shape of a segment of a straight line; the same occurs in the four triangles *KLP*, *LMQ*, *MKN*, *NPQ*.

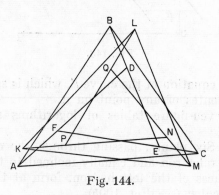

Fig. 144.

In addition, any of the triangles of the first four overlaps partially with any triangle of the second four (and conversely, of course).

Therefore, it we take a point *S* above the plane of the drawing, and a point *T* below the plane of the drawing, then any two of the eight tetrahedra *SABD*, *SBCE*, *SCAF*, *SDEF*, *TKLP*, *TLMQ*, *TMKN*, *TNPQ* have a common boundary in the form of a certain portion of a face.

(74) If we take into account all n_1 paths issuing from an "n_1-node", all n_2 paths issuing from an "n_2-node", etc., and finally, all n_s paths issuing from an "n_s-node", then each path is accounted for twice; therefore, the total number of paths equals $\dfrac{n_1 + n_2 + \ldots + n_s}{2}$.

If the numerator of this fraction were to contain an odd number of odd terms, the total number of paths would be fractional!

(75) In Fig. 118, seven points are broken up into two groups: K, L, M, N and A, B, C. The arrangement of paths is such that from any point of the first group it is possible to go over directly only to a particular point of the second group and vice versa.

In order to visit all points, we have to begin the journey at some point of the first group and complete it at some point of the same group; but is impossible to get from the last point to the initial point directly.

(76) If we inscribe in a regular octahedron a cube with vertices at the centres of the faces of the octahedron, a move from one face of the octahedron to another is possible only when the vertices of the cube lying in these faces are also the ends of some edge of the cube.

Similar arguments apply also in the case of a dodecahedron inscribed in a regular icosahedron.

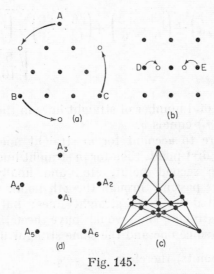

Fig. 145.

(77) In Fig. 145a the arrows show how the coins A, B, C should be moved.

(78) In Fig. 145*b* the arrows show where to put coins *D* and *E*.

(79) The solution is shown in Fig. 145*c*.

(80) The points should be placed at the vertices and at the centre of a regular pentagon (see Fig. 145*d*.)

(81) To the six points of the preceding problem we must add two points, A_7 and A_8, situated on the perpendicular to the plane of the drawing through the centre A_1, and the following should hold: $A_1 A_7 = A_1 A_8 = A_1 A_2$ (although one of the "triangles" $A_1 A_7 A_8$ has an angle of 180° at a vertex).

(82) It is sufficient to prove that the length of segments *OA, OB, OC, OD, AB, AC, AD, BC, BD, CD* are equal. For example

$$AC = \sqrt{\left(\frac{1}{2}-1\right)^2 + \left(\frac{\sqrt{3}}{6}-0\right)^2 + \left(\frac{\sqrt{6}}{3}-0\right)^2 + (0-0)^2} =$$

$$= \sqrt{\frac{1}{4}+\frac{3}{36}+\frac{6}{9}} = 1,$$

$$CD = \sqrt{\left(\frac{1}{2}-\frac{1}{2}\right)^2 + \left(\frac{\sqrt{3}}{6}-\frac{\sqrt{3}}{6}\right)^2 + \left(\frac{\sqrt{6}}{12}-\frac{\sqrt{6}}{3}\right)^2 + \left(\frac{\sqrt{10}}{4}-0\right)^2} =$$

$$= \sqrt{\frac{6}{16}+\frac{10}{16}} = 1$$

etc.

(83) The total number of straight lines in the configuration (p_m, q_n) equals *q*.

If we were to account for *m* straight lines passing through the first point, then for *m* straight lines passing through the second point, etc., and finally, for *m* straight lines passing through the *p*th point, we would have "counted up" *pm* straight lines; but, in this case, each straight line would have been accounted for *n* times (since one and the same straight line passes through *n* points) therefore $\frac{pm}{n} = q$.

(84) We should note, that, whatever Ivanov is, he must answer the question for the teacher, whether he

is a serious one or a joker, in one way only: "I am serious". But then it is clear that Petrov is serious, and Sidorov is a joker.

(85) Since of the numbers 26, 27, 28, only 27 is divisible by 3, therefore Galkin and Komkov were sawing up logs of length $1\frac{1}{2}$ m, and therefore their names are Petya and Kostya. But Kostya was not among the team leaders, and it was given that Komkov was not a team leader. Therefore Komkov is Kostya.

(86) Yes, he could. Indeed, from Andrey's answer, it should be clear to his friends that they cannot be wearing two white hats.

If Vadim were wearing a white hat, Boris could have determined the colour of his hat easily, but since he could not do that, Vadim's hat must have been black. The problem can be generalized easily: we can speak of n friends, sitting one behind the other, and of $2n + 1$ hats (n white ones and $n + 1$ black ones).

(87) Assuming that Seryozha is No. 2, (i. e. he took second place) we have: From 1 — Kolya is not No. 3, from IV — Vanya is not No. 4, from V — Kolya is No. 1, from II — Tolya is not No. 1, Nadya is No. 2, i. e. we arrive at a contradiction: both Seryozha and Nadya took second place.

Assuming, that Kolya is No. 3, we obtain from I — Seryozha is not No. 2, from IV — Vanya is No. 4, from II — Nadya is not No. 3, Tolya is No. 5, from III — Tolya is not No. 1, Nadya is No. 2: therefore Seryozha's lot is place No. 1. Thus in the order of their numbers we have; Seryozha, Nadya, Kolya, Vanya, Tolya.

(88) Since it is known that three of the chess-players hail from Saratov, Moscow and Kiev, and their ages are 21, 22 and 23 years, therefore the fourth chess-player comes from Fergana and he is 20 (student M); he is a biologist (since the remaining three chess players are a mathematician, a chemist, a geologist) and he is in his fourth year of study (the remaining chess-players

are 1st, 2nd and 3rd year students). These data are entered under the numbers 1, 2, 3 in the table, where the data given in the problem are entered in bold print.

Town \ Age	20 years	21 years	22 years	23 years
Kiev	A. geol. (5) 2nd, footb. (17) (28)	B. chem. (4) 4th, box. (18) (29)	C. biol. 3rd, volleyb. (22) (30)	D. math. 1st, chess
Moscow	E. math. (6) 3rd, box. (16) (27)	F. biol. (13) 1st, footb. (19)	G. chem. 2nd, chess	H. geol. (12) 4th, volleyb. (26) (31)
Saratov	I. chem. (7) 1st, volleyb. (15)	J. geol. 3rd, chess	K. math. (8) 4th, footb. (21) (33)	L. biol. (9) 2nd, box. (25) (34)
Fergana	M. biol. (2) 4th, chess (3) (1)	N. math. (14) 2nd, volleyb (20) (32)	O. geol. (10) 1st, box. (23) (35)	P. chem. (11) 3rd, footb. (24) (36)

Under No. 4 we write down, that *B* is a chemist (since the Kiev men, *C* and *D* are a biologist and a mathematician respectively, and the student *J* is 21 years old and he is a geologist).

Under the numbers 5–14, easily determinable information about the specialization of the remaining students is entered in the table. Then, under number 15, we note down: the chemist from Saratov, *I*, is a 1st year student (since the Saratov man *J* is a 3rd year student, the chemist *G* is a second year student, and the

twenty-year-old *M* is a fourth year student). The
courses followed by the remaining students are determined similarly (under 16–26).

The twenty-year old Muscovite *E*, is a boxer (No. 27)
since the Muscovites *F* and *G* are a footballer and a chess
player respectively, and the twenty-year old *I* is a volleyball player. Further, under numbers 28–36, the favourite forms of sport of the students are indicated (for
example, No. 29 *B* is a boxer, since the Kiev men
A and *D* are a footballer and a chess player respectively,
and the chemist *I* is a volleyball player and so on).

(89) (1)

$$\sqrt{\text{******}} = \text{***}$$

```
      ****** = ***
     _*
      ___
      ***
    _ **
      ____
      4***
      ****
      ____
       0
```

It is seen immediately, that the first digit of the root
required is 3, since its square is a one-digit number
and it is a result of the extraction of square roots from
the upper division consisting of two digits.

The second digit of the root required can only be
unity, since even 62×2 would have given a three-
digit product, and not a two-digit one. (See two asterisks
in the fourth row). Finally, it follows from $62z \times z =
= 4\text{***}$, that *z* (the last digit of the root sought) can
only be a seven.

```
2)   ******  |***
    _****     |*8*
     ___
     ***
    _***
     ____
     ****
    _****
     ____
      0
```

It may be seen immediately that the quotient is 989, since the product of the divisor and 8 is a three-digit number, and its product with the end numbers of the quotient are four-digit numbers (see the 4th, 2nd and 6th rows of the pattern of division).

Let us denote the whole of the divisor by z. Then $8z < 1000$ and $9z \geqslant 1000$, i.e. $111 < z < 125$. When $z = 112$, the dividend $= 989 \times 112 = 110\,768$ and the division of the number $110\,768$ by 112 takes place in exact accordance with the given pattern. But for $z = -113, 114$, etc., the first remainder would not have been a two-digit one as it would follow from the pattern, but a three-digit one; for example, when $z = 113$, we have

$$\begin{array}{l} 111757 (= 989 \times 113). \\ \underline{1007 \quad (= 113 \times 9)} \\ 100 \end{array} \qquad \begin{array}{l} \text{when } z = 114: \; 112746 \; (989 \times 114) \\ \underline{\qquad = 1026 \quad (= 114 \times 9)} \\ 101 \end{array}$$

etc. Thus the only solution is $110\,768 - 112 - 989$.

(3)
$$\begin{array}{l} \text{s m e h} \\ \underline{\text{g r o m}} \\ \text{g r e m i} \end{array}$$

It can be seen immediately, that $g = 1$. Therefore $p \neq 1$; but p cannot be greater than one, either, since when $p = 2$, even if $c = 9$, on adding up the hundreds, there should be "carry two" and that is impossible. Thus $p = 0$.

It is easy to see that $m < 9$, since if $m = 9$, on investigating addition in the order of hundreds, we would have obtained $e = 0$, and this is impossible, since $p = 0$. Therefore, $m < 9$ and $e = m + 1$.

Evidently, $c = 9$. On investigating addition in the order of tens, we obtain: $o = 8$, since $e = m + 1$ and $o \neq c = 9$.

So, we have

$$\begin{array}{l} \text{9 m e h} \\ \underline{\text{1 0 8 m}} \\ \text{1 0 e m i} \end{array}$$

By investigating five variants: $m = 2, 3, 4, 5, 6$, it is easy to discover that $m = 5$ is the only one that does not lead to a contradiction, but gives: $9567 + 1085 = 10\,652$.

4) *forty*
    ten
  + *ten*
  ——————
    sixty

It is easy to establish that $n = 0$ and $e = 5$. Since $i \neq 0$, therefore $i = 1$ (if $i = 2$, even if $o = 9$, the addition in the order of hundreds would collect "carry three", which is impossible). For the same reason, $o = 9$. Thus we have:

      *f*9*rty*
       *t*50
    + *t*50
    ——————
      s1xty

Therefore, $s = f + 1$ and in the order of hundreds we collect "two to carry". But for $f = 7$ and $s = 8$, it is possible to get in the order of hundreds: $8 + 8 + 4 + 1 = 21$, or $8 + 8 + 3 + 1 = 20$, but then x coincides either with i or with n.

It is easy to verify that when $f = 3$ and $s = 4$, x coincides with one of the "digits" s, f, i, n, in all variants.

Only when $f = 2$, $s = 3$, and $r = 7$, $t = 8$, do we get a value for x, that does not coincide with any one of the other "digits": $x = 4$, and then $y = 6$.

Thus, the only solution is $29\,786 + 850 + 850 = 21\,486$.

(90) (1) Cross: Aa AB ab

Return:	A	a

(2) Cross: abc AB BCD abc cd

Return:	c	Bb	a	c

(3) Cross: abc cde ABC CDE bed de

Return:	c	de	ce	b	d

(91) The first method gives $(20, 0, 0) \rightarrow (7, 13, 0) \rightarrow (7, 4, 9) \rightarrow (16, 4, 0) \rightarrow (16, 0, 4) \rightarrow (3, 13, 4) \rightarrow (3, 8, 9) \rightarrow (12, 8, 0) \rightarrow (12, 0, 8)$. As we have arrived at the situation $(b - 1, 0, c - 1)$, the first method is unsuitable.

The second method gives $(20, 0, 0) \rightarrow (11, 0, 9) \rightarrow (11, 9, 0) \rightarrow (2, 9, 9) \rightarrow (2, 13, 5) \rightarrow (15, 0, 5) \rightarrow (15, 5, 10) \rightarrow (6, 5, 9) \rightarrow (6, 13, 1) \rightarrow (19, 0, 1) \rightarrow (19, 1, 0) \rightarrow (10, 1, 9) \rightarrow (10, 10, 0)$, i. e. it leads to the required goal.

(92) The first method gives $(16, 0, 0) \rightarrow (4, 12, 0) \rightarrow (4, 5, 7) \rightarrow (11, 5, 0) \rightarrow (11, 0, 5)$; it is unsuitable as it leads to the position $(b \rightarrow 1, 0, c \rightarrow 2)$. The second method gives $(16, 0, 0) \rightarrow (9, 0, 7) \rightarrow (9, 7, 0) \rightarrow (2, 7, 7) \rightarrow (2, 12, 2) \rightarrow (14, 0, 2) \rightarrow (14, 2, 0) \rightarrow (7, 2, 7) \rightarrow (7, 9, 0) \rightarrow (0, 9, 7) \rightarrow (0, 12, 4) \rightarrow (12, 0, 4) \rightarrow (12, 4, 0), \rightarrow (5, 4, 7) \rightarrow (5, 11, 0)$. As we have arrived at a situation, where the small vessel cannot be filled (from the large vessel) and a portion equal to the capacity of the medium vessel cannot be returned to the large vessel, the second method is also unsuitable for dividing the liquid into two equal parts.

(93) A side of the base of a circumscribed pyramid equals $2n + 1$, and its height equals $n + \frac{1}{2}$ (if the central cubes of all layers are in one column). Therefore

$$1^2 + 3^2 + 5^2 + \ldots + (2n-3)^2 + (2n-1)^2 =$$

$$= \frac{(2n+1)^2 \left(n + \frac{1}{2} \right)}{3} - 2(2n-1) + (2n-3) + \ldots$$

$$\ldots + 5 + 3 + 1] - 4n \cdot \frac{1}{3} - \frac{1}{6} = \frac{n(4n^2 - 1)}{3}.$$

(94) There are 31 squares, 124 triangles, 87 rectangles (including the squares).

(95) Thirty-five triangles.

(96) 78 triangles, 11 regular hexagons, 66 rhombi.

(97) (1) Suppose $N_1 (m, n)$ is the number of squares, and $N_2(m, n)$ is the number of rectangles which can be seen on an "m, n-board".

If $m \leq n$ and a is the side of a square of a chessboard, then an "m, n-board" contains m columns of width a, each containing n squares of side a, $n - 1$ columns of width $2a$, each containing $n - 1$ squares of side $2a$, $m - 2$ columns of width $3a$, each containing $n - 2$ squares of side $3a$, etc., finally, one column of width ma, containing $n - m + 1$ squares of side ma. Therefore

$$N_1(m, n) = mn + (m-1)(n-1) + (m-2)(n-2) + \dots$$
$$\dots + 2(n-m+2) + 1(n-m+1).$$

It is possible to select a column of any particular width on an "m, n-board" in $\frac{m(m+1)}{2}$ ways. Having selected a column it is possible to see in it $\frac{n(n+1)}{2}$ rectangles with base equal to the width of the column (n rectangles of altitude a, $n - 1$ rectangles of altitude $2a$, etc., finally, two rectangles of altitude $(n - 1) a$ and one of altitude na).

Therefore, $N_2(m, n) = \frac{m(m + 1)}{2} + \frac{n(n + 1)}{2}$. In particular $N_1(n, n) = n^2 + (n - 1)^2 + (n - 2)^2 + \dots + 3^2 + 2^2 + 1^2 = \frac{n(2n + 1)(n + 1)}{6}$: $N_1(8,8) = 204$; $N_2(n, n) = \frac{n^7(n + 1)^2}{4}$; $N_2(8, 8) = 1296$.

(2) We can select a square of side k cm in the base of a "10^3-cube" in $(10 - k + 1)^2$ ways. In the column with the selected base there are $(10 - k + 1)$ "k^3-cubes". Therefore, in a "10^3-cube" it is possible to "see" $(10 - k + 1)^3$ "k^3-cubes". Since k can be any number from 1 to 10, therefore it is possible to see in a "10^3-cube" 3025 cubes of various sorts ($1^3 + 2^3 + 3^3 + 4^3 + 5^3 + 6^3 + 7^3 + 8^3 + 9^3 + 10^3$).

Since a rectangle can be selected in the base of a "10^3-cube" in 3025 ways ($= N_2(10, 10) = 55^2$), and in a column with any particular base we can see 55 rectangular parallelepipeds with that base $(10 + 9 + 8 + \ldots + 2 + 1)$, therefore it is possible to see 55^3 different rectangular parallelepipeds in a "10^3-cube".

(98) In forty-four ways.

(99) Adding the following identities term by term

$$\begin{cases} \sin \alpha \, \sin \dfrac{\alpha}{2} = \dfrac{1}{2} \cos \dfrac{\alpha}{2} - \dfrac{1}{2} \cos \dfrac{3\alpha}{3}, \\[2mm] \sin 2\alpha \, \sin \dfrac{\alpha}{2} = \dfrac{1}{2} \cos \dfrac{3\alpha}{2} - \dfrac{1}{2} \cos \dfrac{5\alpha}{2}, \\[2mm] \cdots \cdots \cdots \cdots \cdots \cdots \\[2mm] \sin n\alpha \, \sin \dfrac{\alpha}{2} = \dfrac{1}{2} \cos \dfrac{(2n-1)\alpha}{2} - \dfrac{1}{2} \cos \dfrac{(2n+1)\alpha}{2}, \end{cases}$$

we obtain

$$(\sin \alpha + \sin 2\alpha + \ldots + \sin n\alpha) \sin \dfrac{\alpha}{6} =$$

$$= \dfrac{1}{2} \cos \dfrac{\alpha}{2} = \dfrac{1}{2} \cos \dfrac{(2n+1)\alpha}{2} = \sin \dfrac{(n+1)\,\alpha}{2} \sin \dfrac{n\alpha}{2}.$$

whence

$$\sin \alpha + \sin 2\alpha + \ldots + \sin n\alpha = \dfrac{\sin \dfrac{(n+1)\alpha}{2} \sin \dfrac{n\alpha}{2}}{\sin \dfrac{\alpha}{2}}.$$

(100)

$$[3(10^k + 10^{k-1} + \ldots + 10 + 1)n + 1]^2 =$$

$$= \left[\dfrac{n(10^{k+1} - 1)}{3} + 1 \right]^2 = \dfrac{n^2(10^{2k+2} - 2 \times 10^{k+1} + 1)}{9} +$$

$$+ \dfrac{2n(10^{k+1} - 1)}{3} + 1 = n^2 \dfrac{10^{2k+2} - 10^{k+1}}{10 - 1} +$$

$$(6n - n^2) \dfrac{10^{k+1} - 1}{10 - 1} + 1 = n^2(10^{2k+1} + 10^{2k} + \ldots +$$

$$+ 10^{k+1}) + (6n - n^2)(10^k + 10^{k-1} + \ldots + 1) + 1.$$

(101) Since the volume of the liquid in the first tumbler remains unchanged after two pourings, therefore the volume of spirits which passed into the first tumbler equals the volume of water which passed into the second tumbler (the note about "thorough mixing would have needed to be used in the arithmetical — much longer! — solution, which, true, would have enabled us to determine also the amount of admixture in each tumbler).

(102) The portrait on the wall was that of Petrov's grandson. (Petrov's fancy answer could be rephrased thus; "I am the grandfather of the hanging one").

(103) Since everyone has 8 great-great-grandfathers and 8 great-great-grandmothers, and each of these sixteen persons had 16 direct ancestors in the "fourth generation" in his turn, the number required is 256 (16×16).

The number of direct ancestors in the "eighth generation" would be less than 256 if there were cases, say of marriages between second cousins, etc.

(104) $15'$; if we join up any points A and C situated on the sides of the angle ABC, we see through a magnifying glass a similar triangle $A'B'C'$ with angles equal to those of triangle ABC.

(105) By 25%. (If say, one kg of potatoes costs a roubles, then its cost after price reductions is $0.8a$ roubles, and a roubles now buys 1.25 kg potatoes.)

(105a) By 50%. (If before reductions, 100 roubles buys b kg of potatoes, 100 roubles could buy $1.2b$ kg of potatoes after the first reduction, and $1.2 \times 1.25b = 1.5\ b$ kgm of potatoes after the second reduction.)

(106) 44 times. (In 24 hr the minute hand makes 24 revolutions and the hour hand makes 2 revolutions; therefore, the minute hand overtakes the hour hand 22 times, and the hands form a right angle twice between each two "overtakings".)

(107) Suppose at present Kolya is x years old and Olya is y years old. It follows from the conditions of the problem that aunt Polya was $x + y$ years old at the

time when Kolya was y years old, i. e. aunt Polya is x years older than Kolya, so that she is $2x$ years old at present; y years ago she was "of Kolya's age" and Kolya was a newly born baby.

(108) The pilot saw a white (polar) bear, because it follows from the condition $AB = AC$, that the point A is at the North Pole. (A could be also near the South Pole — see note (121) — but there are no bears there.)

(109) Denote the number of oxen by $10a + b$ ($0 \leqslant b \leqslant 9$). A and B gained $100a^2 + 20ab + b^2$ roubles from the sale. The part of the sum gained $100a^2 + 2ab$ was distributed equally by taking 10 roubles each in turn. In the remaining amount (b^2 roubles) the number of tens must be odd, because when A took the last 10 roubles in his turn, there remained less than ten roubles.

Since, when $0 \leqslant b \leqslant 9$, $b^2 = 0, 1, 4, 9, 16, 25, 36, 49, 64, 81$ and only in the numbers 16 and 36 the number of tens is odd, the remainder turns out to be 6 roubles. From the equation $10 - x - 6 + x$ we find that the purse cost 2 roubles.

(110) Suppose the husbands bought s, x, z objects, and their wives bought t, y and w objects respectively. Then $s^2 - t^2 = x^2 - y^2 = z^2 - w^2 = 45$, or $(s + t)(s - t) = (x + y)(x - y) = (z + w)(z - w) = 45$.

But the number 45 can be factorized in three ways only: $45 = 45 \times 1 = 15 \times 3 = 9 \times 5$. From the equations $s + t = 45$ and $s - t = 1$ we find $s = 23$ and $t = 22$. Similarly, we find $x = 9$, $y = 6$ and $z = 7$, $w = 2$. Therefore, the husbands spent: 529 roubles (Yuri), 81 roubles (Alexander) 49 roubles (Login), and their wives spent: 484 roubles (Tatyana), 36 roubles (Nina), 4 roubles (Olga).

(111) Suppose the total amount of money in the general case is x (when $\frac{1}{n + 1}$ part of the remainder is added each time).

Then $1 + \dfrac{x - 1}{n + 1} = 2 + \dfrac{x - 3 - \dfrac{x - 1}{n + 1}}{n + 1}$, whence $x = n^2$

The first child's share is (in roubles): $1 + \dfrac{n^2-1}{n+1} = n$,

the second child's share is $2 + \dfrac{n^2-n-2}{n+1} = 2 + (n-2) =$

$= n$, the third child's share is $3 + \dfrac{n^2-2n-3}{n+1} = 3 + (n-3) =$

$= n$, etc. If we suppose that the first k children received n roubles each, the $(k+1)$th child's share is

$k + 1 \dfrac{n^2-kn-(k+1)}{n+1} = k + 1 + n - (k+1) = n$. It

follows hence, that each child received n roubles.

(112) If Gleb walked x km (towards the end of the road), then Pavel walked the same distance at the beginning of the road (they arrived at N simultaneously!)

While Gleb walked x km with a speed of u km/hr, Yuri travelled $2s - 3x$ km with a speed of v km/hr ($s - 2x$ to meet Pavel and $s - x$ to catch up with Gleb). Therefore $\dfrac{x}{2s-3x} = \dfrac{u}{v}$, whence $x = \dfrac{2su}{v+3u}$

and the time taken was $\dfrac{u}{x} + \dfrac{s-x}{v} = \dfrac{s(u+3v)}{v(v+3u)}$ hr.

(113) The circumference of the clock face is divided into 60 minute "divisions". Suppose the job began at 4^x o'clock, and ended at 7^y o'clock. As the hour hand moves 12 times slower than the minute hand, we have

two equations: $\begin{cases} x = 12(y-20) \\ y = 12(x-35) \end{cases}$, whence $x = \dfrac{5280}{143} = x_0$,

$y = \dfrac{3300}{143} = y_0$.

It is easy to verify that $x_0 - 30 = 30 - y_0$ (see Fig. 146).

(114) If the time shown by the clock retains sense, when the hour hand and the minute hand change places in positions "m hr x min" and "n hr y min", we must

have $\begin{cases} x = 12(y - 5m) \\ y = 12(x - 5n) \end{cases}$, whence $x = \dfrac{60(m+12n)}{143}$,

$y = \dfrac{60(n+12)m}{143}$.

Since m and n vary from 0 to 11 inclusively, each pair of numbers m, n ($m \neq n$) gives two instants when the hands can interchange places "painlessly". In 12 hr there are 132 ($2 \times C^2_{12}$) such instants, and in 24 hr there are 264 such instants.

In addition, the hour hand and the minute hand "coincide" 22 times in 25 hr, i. e. they can "interchange places" without changing positions.

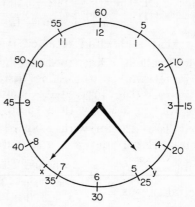

Fig. 146.

(115) Every s-digit number n is confined within the limits $10^{s-1} \leq n < 10^s$. There are altogether $9 \times 10^{s-1}$ s-digit numbers and the total number of digits in them is $9s \times 10^{s-1}$. In all s-digit numbers the first digit is not zero, and in the remaining places zero and one of the other digits is encountered an equal number of times. Therefore, the number of zeros in all s-digit numbers equals $\dfrac{9s \times 10^{s-1} - \times 10^{s-1}}{10} = 9(s - 1) \times 10^{s-2}$ (the same number as that of all digits in all $(s - 1)$ digit numbers).

If N_1 is the total number of digits in the sequence of numbers 1, 2, 3, ..., $10^k - 1$, 10^k and N_2 is the total number of zeros in the sequence 1, 2, 3, ..., $10^{k+1} - 1$, 10^{k+1}, then $N_1 = N_2 = 9 + 9 \times 2 \times 10 + 9 \times 3 \times \times 10^2 + \ldots + 9 \times k \times 10^{k-1} + k + 1$ ($k + 1$ is the number of digits in the number 10^k, or the number of zeros in 10^{k+1}).

(116) We place the figure S on paper in any way desired (Fig. 147). We then transfer in a parallel manner all squares occupied by the figure into some single square w. Here the areas occupied by the figure may overlap.

In w there is bound to be a point A not occupied by pieces of figure S. There are points A_1, A_2, ... in other squares, which occupy the same positions in other squares as A occupies in w, and are not covered by portions of S. It is sufficient to shift the paper under S, so that the nodes coincide with A, A_1, A_2, etc.

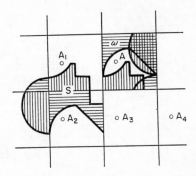

Fig. 147.

(117) If we are given a point A and two straight lines l_1 and l_2 distributed at random in space, it is, generally speaking, possible to draw a straight line m through A, which intersects l_1 and l_2 (this is impossible in exceptional cases, when $l_1 \parallel l_2$ and A does not lie in the plane containing l_1 and l_2). It is sufficient to pass planes through A and l_1 and through A and l_2 and their intersection is the required straight line m.

By picking various points A on the straight line l_3 we obtain an infinite set of lines m_1, m_2, . . ., each intersecting l_1, l_2, l_3.

The straight lines m_1, m_2, m_3, . . . form a so-called ruled surface, which is intersected by l_4, generally speaking, at some point B lying on some straight line m_1. This straight line intersects l_1, l_2, l_3, l_4.

(118) Suppose R_s is the radius of the circle C_s ($0 \leq \leq s \leq 1000$), 0 is the centre of circle C_0 and O_m is the centre of circle C_m ($1 \leq m \leq 1000$).

The radii of circles C_0, C_k, C_{k+1} are connected by the relationship

291

$$R_{k+1} = \frac{R_0 R_k}{(\sqrt{R_0} - \sqrt{R_k})^2} ; \qquad (1)$$

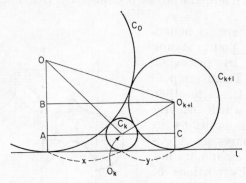

Fig. 148.

(from the triangle OO_kA (see Fig. 148) we have $(R_0 + R_k)^2 - (R_0 = R_k)^2 = x^2$ or $x = 2\sqrt{(R_0 R_k)}$; similarly, from the triangles $OO_{k+1}B$ and $O_kO_{k+1}C$ we get $(x + y) = 2\sqrt{(R_0 R_{k+1})}$ and $y = 2\sqrt{(R_k R_{k+1})}$; therefore

$$\sqrt{R_0 R_{k+1}} - \sqrt{R_k R_{k+1}} = \sqrt{R_0 R_k} .$$

Expressing the lengths of radii in km we have $R_0 - 1$, $R_1 = 10^{-6}$. According to formula (1), for $k = 1$ we get

$$R_2 = \frac{1 \times 10^{-6}}{(1 - 10^{-3})^2} = \frac{1}{999^2},$$

and for $k = 2$,

$$R^3 = \frac{1}{999^2} \times \left(1 - \frac{1}{999}\right)^2 = \frac{1}{998^2}.$$

Assuming that

$$R_m = \frac{1}{(1000 - m + 1)^2}, \qquad (2)$$

from formula (1) we get $R_{m+1} = \dfrac{1}{(1000 - m)^2}$ (transition from m to $m + 1$) and, as formula (2) holds for m

equal to unity, it is true for any m not exceeding one thousand, since, when $m = 1000$, it gives $R_{1000} = 1$ and the circle C_{1001} cannot be constructed.

(119) Let the direction of the ray m, incident on the plane Oxy be characterized by the unit vector $e = \overline{AO} = (e_x, e_y, e_z)$ (the brackets contain the components of vector e parallel to the coordinate axes). The ray m', reflected from the plane Oxy (or from a plane, parallel to Oxy) has, as unit vector, $e_1 = (e_x, e_y, -e_z)$, the only change is the sign of the components parallel to the axis O_z (see Fig. 149).

The unit vectors of rays m'' and m''', obtained on reflecting ray m' from the plane parallel to plane Oxz and of ray m'' from the plane parallel to the plane Oyz are $e_2 = (e_x, -e_y, -e_z)$.

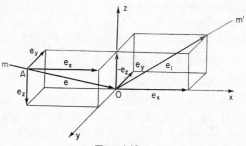

Fig. 149.

and $e_3 = (-e_x, -e_y, -e_z)$. Therefore, $m''' \parallel m$ and the direction of the ray m''' is opposite to that of ray m.

(120) The components α, β, γ (parallel to the edges of the parallelepiped) of the unit vector $e\,(\alpha,\ \beta,\ \gamma)$ characterizing the required direction of the ray, do not alter in absolute magnitude on being reflected from the faces of the parallelepiped. Therefore, if an imaginary point moves along the incident ray and along all reflected rays, the sums of its displacements in the direction of each of the axes Ox, Oy and Oz (regarding as positive the displacements both up and down, rightwards and leftwards and "towards us" and "away from us") are proportional to the numbers $|\alpha|$, $|\beta|$, $|\gamma|$.

But, at the time of the point's return, after reflection from all faces, to the starting point, these summary displacements equal the doubled lengths of edges, a, b, c of the parallelepiped, therefore $|\alpha| : |\beta| : |\gamma| = a:b:c$, i. e. the ray should be directed parallel to one of the diagonals of the parallelepiped.

The solution of the problem does not really become any more complicated, if the ray sought is to be reflected from the left face, the right face and the back face three times, from the front face twice, from the lower face four times and from the upper face five times. Try to give the corresponding arguments when the distances of the starting point from the left, front and bottom faces of the parallelepiped are a', b', c' respectively.

(121) The obvious solution: A — North Pole; less obvious solutions: A — any point in the Southern Hemisphere, situated on the parallel l' 2000 km to the North of parallel l, whose length equals $\frac{2000}{n}$ km (n is any natural number).

(122) Since the aeroplane flew a km eastwards and found itself at the end of the journey $3a$ km east of Leningrad, it must have been flying eastwards along a "parallel φ" whose radius r is three times smaller than the radius r of the Leningrad parallel (latitude $60°$ North); therefore, $R\cos = \varphi\, r_1 = \frac{r}{3} = \frac{R\cos 60°}{3} = \frac{R}{6}$,

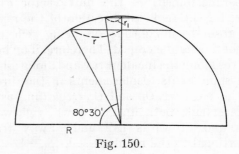

Fig. 150.

whence $\varphi \approx 80°30'$ (Fig. 150 gives the projection of the Northern Hemisphere on to a plane parallel to the earth's axis). Therefore a is the length of the arc of the meridian between parallel 60° and parallel 80°30' i. e. $a \simeq \dfrac{40000 \text{ km} \times 20.5}{360} \simeq 2278$ km.

(123) It should be taken into account that the sum of angles of a spherical triangle is always greater, than 180°, and its area is calculated from the formula

$$S_{ABC} = (\alpha + \beta + \gamma - \pi)R^2$$

where α, β, γ are the values of the angles A, B, C expressed in radian measure.

Indeed, if any pair of the sides of the spherical triangle BC are produced to meet (see Fig. 151) we get

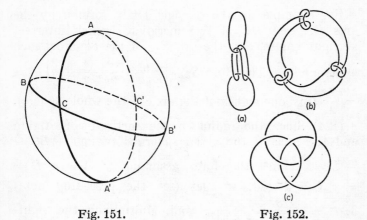

Fig. 151.   Fig. 152.

"lunes" $ABA'CA$, $BAB'CB$, $CAC'BC$, the areas of whose surfaces are

$$S_{ABA'CA} = 4\pi R^2 \cdot \frac{\alpha}{2\pi}, \tag{1}$$

$$S_{BAB'CB} = 4\pi R^2 \cdot \frac{\beta}{2\pi} \tag{2}$$

and

$$S_{CAC'BC} = S_{ABC} + S_{ABC'} = 4\pi R^2 \cdot \frac{\gamma}{2\pi}.$$

If we substitute $S_{A'B'C}$ for $S_{ABC'}$, in the last equation, we get

$$S_{ABC} + S_{A'B'C} = 4\pi R^2 \cdot \frac{\gamma}{2\pi} \qquad (3)$$

(any two sides of spherical triangles ABC' and $A'B'C$ are alike, and although these triangles cannot, generally speaking, coincide on being superimposed upon each other, it can be proved that $S_{A'B'C} = S_{ABC'}$).

Summing eqns. (1), (2) and (3) term by term we get

$$2S_{ABC} = 2\pi R^2 - 2R^2(\alpha\beta + +\gamma),$$

whence

$$S_{ABC} = (\alpha + \beta + \gamma - \pi) R^2.$$

For example, in the triangle DEF, formed by the arcs DE and DF of two meridians, which intersect at right angles, and the arc EF of the equator, each angle equals $\frac{\pi}{2}$. Therefore, $S_{DEF} = \left(\frac{\pi}{2} + \frac{\pi}{2} + \frac{\pi}{2} - \pi\right) R^2 = \frac{\pi R^2}{2}$ (just one eighth of the area of the whole sphere).

(124) Since the radius of the earth $R \simeq 6370$ km and the area of the given spherical triangle $ABC - S \simeq \frac{\sqrt{3}}{4}$ km², therefore, assuming $< A = < B = = < C = \alpha$ rad, we get (see the preceding note) $3\alpha - \pi = \frac{S}{R^2} \simeq \frac{\sqrt{3}}{4 \times 6370^2}$ rad. Multiplying the right-hand side by $\frac{180 \times 60 \times 60}{\pi}$ in order to express it in seconds, we obtain $< A = < B = < C \simeq 60°0'0.0007''$

(125) Figure 152 shows three methods of linking three string rings; the first two methods can be easily used also for linking n rings in such a way that if one ring is broken the remaining rings can be parted without any further breakages.

Bibliography

[1] ARENS, V. (1924) *Mathematical Games and Pastimes.*

[2] ARNOLD, I. V. (1939) *Theory of Numbers.*

[3] BARLOW (1958) *Tables of Squares, Cubes, Square Roots, Cube Roots and Reciprocals of All Numbers, Up to* 12500, Spon, London.

[4] BERMAN, G. N. (1948) *The Cycloid.*

[5] BOLTYANSKII, V. G. (1956) *Equal and Equi-composed Figures.*

[6] BOBROV, S. (1948) *The Magic Two-Horn.*

[7] VOROB'EV, N. N. (1951) *Fibonacci Numbers.* (English Translation: Pergamon Press, 1961)

[8] HILBERT, D. and KOHN-VOSSEN, S. (1951) *Anschauliche Geometrie.*

[9] DOMORYAD, A. P. (1951) *Numerical and Graphic Methods of Solving Equations* Encyclopaedia of Elementary Maths., Book II.

[10] DOMORYAD, A. P. (1959) *Computing Appliances* Children's Encyclopaedia, vol. 3.

[11] DOMORYAD, A. P. (1957) *On Calculating Logarithms* Proceedings of the Tashkent State Pedagogical Institute, VII issue.

[12] ZHITOMIRSKII, O. K. *et al.* (1935) *Text Book in Higher Geometry*, Part I.

[13] KORDEMSKII, B. A. (1954) *Mathematical Intuition.*

[14] KORDEMSKII, B. A., and RUSALYOV, N. V. (1952) *The Amazing Square.*

[15] LITZMAN, V. (1959) *Giants and Dwarfs in the World of Numbers.*

[16] LURYE, S. Y. (1945) *Archimedes.*

[17] MARKUSHEVICH, A. I. (1951) *Recurring Sequences.*

[18] MÄNNCHEN, F. (1923) *Some Secrets of Performing Calculators.*

Bibliography

[19] OBREIMOV, V. I. (1884) *A Triple Puzzle.*

[20] OKUNEV, L. YA. (1949) *Higher Algebra.*

[21] OKUNEV, L. YA. (1935) *Combinatorial Problems on a Chess-board.*

[22] PEREL'MAN, Y. I. *Entertaining Algebra.* (English Translation: Pergamon Press)

[23] PEREPELKIN, D. I. *A Course of Elementary Geometry,* part I.

[24] RADEMACHER, G. and TEPLITZ, O. (1936) *Numbers and Figures.*

[25] USPENSKII, Y. (1924) *Selected Mathematical Pastimes.*

[26] FIKHTENGOL'TS, G. M. (1948) *A Course of Differential and Integral Calculus,* vol. II.

[27] FIKHTENGOL'TS, G. M. (1931) *Mathematics for Engineer,* vol. I.

[28] KHINCHIN, A. Y. (1935) *Continued Fractions.*

[29] SHKLYARSKII, D. O. *et al.* (1950) *Selected Problems and Theorems of Elementary Mathematics,* part I — *Arithmetic and Algebra.*

[30] SCHUBERT, G. (1923) *Mathematical Games and Pastimes.*

[31] SHUBRIKOV, A. V. (1940) *Symmetry.*

[32] YAGLOM, A. M. and YAGLOM, I. M. (1954) *Non-Elementary Problems in an Elementary Treatment.*

[33] AHRENS, W. *Mathematische Unterhaltungen Bd. I, II.*

[34] AHRENS, W. (1918) *Altes und neues aus der Unterhaltungs-mathematik.*

[35] MITRINOVIC, D. S. (1957) *Sbornik matematickih problema* I.

Journal and Compendia

[36] *Mathematics in Schools.*

[37] *Mathematical Education.*

[38] *Mathematical Enlightenment* (1934—1938).

[39] *American Mathematical Monthly.*